T0092250

Meeting the Challenge

ALSO BY THE AUTHOR

Science in London: A Guide to Memorials (Springer Nature, 2021, with I. Hargittai)

Moscow Scientific: Memorials of a Research Empire (World Scientific, 2019, with I. Hargittai)

New York Scientific: A Culture of Inquiry, Knowledge, and Learning (Oxford University Press, 2017, with I. Hargittai)

Women Scientists: Reflections, Challenges, and Breaking Boundaries (Oxford University Press, 2015)

Budapest Scientific: A Guidebook (Oxford University Press, 2015, with I. Hargittai)

Great Minds: Reflections of 111 Top Scientists (Oxford University Press, 2014, with B. Hargittai and I. Hargittai)

Visual Symmetry (World Scientific, 2009, with I. Hargittai)

Candid Science, vols. 1–6: *Conversations with Famous Scientists* (Imperial College Press, 2000–2006, with B. Hargittai and I. Hargittai)

In Our Own Image: Personal Symmetry in Discovery (Kluwer/Plenum, 2000; Springer, 2012, with I. Hargittai)

Symmetry: A Unifying Concept (Shelter, 1994; Random House, 1996, with I. Hargittai)

Symmetry through the Eyes of a Chemist (3rd edition, Springer, 2009, 2010, with I. Hargittai)

The Molecular Geometry of Coordination Compounds in the Vapor Phase (Elsevier, 1977, with I. Hargittai)

EDITED VOLUMES

Advances in Molecular Structure Research, vols. 1-6 (JAI Press, 1995-2000, with I. Hargittai)

Stereochemical Applications of Gas-Phase Electron Diffraction, Parts A and B (VCH, 1988, with I. Hargittai)

Magdolna Hargittai is a physical chemist and research professor of structural chemistry at the Budapest University of Technology and Economics. She is a member of the Hungarian Academy of Sciences, the Academia Europaea (London), and the European Academy of Sciences (Brussels). She is a PhD and DSc and holds an honorary doctorate from the University of North Carolina. She has been interested in the lives and careers of women scientists and has lectured and published a monograph on this topic, *Women Scientists* (Oxford University Press, 2015). She and her husband, Istvan, have jointly published extensively on symmetry, science history, conversations with famous scientists, and more recently on the memorials of scientists in Budapest, New York, Moscow, and London. Her books have appeared in English, Hungarian, Russian, German, Swedish, and Korean. Among her other distinctions, in 2011, she received the IUPAC "Distinguished Women Chemists and Chemical Engineering Award." Also in 2011, the Hungarian Science Journalists awarded jointly her and her husband the distinction "Educational Scientist of the Year," and named a small planet 2006 HZ17 after them: (192155) Hargittai. The Hargittais live in Budapest. Their son is a professor of chemistry in the United States and their daughter is a professor of communication studies in Switzerland.

Meeting the Challenge

Top Women in Science

MAGDOLNA HARGITTAI

OXFORD
UNIVERSITY PRESS

Oxford University Press is a department of the University of Oxford. It furthers the University's objective of excellence in research, scholarship, and education by publishing worldwide. Oxford is a registered trade mark of Oxford University Press in the UK and certain other countries.

Published in the United States of America by Oxford University Press
198 Madison Avenue, New York, NY 10016, United States of America.

CIP data is on file at the Library of Congress
ISBN 978–0–19–757475–1

DOI: 10.1093/oso/9780197574751.001.0001

Printed by Sheridan Books, Inc., United States of America

Contents

Foreword

As the daughter of a successful scientist mother—the author of this volume—I have lived my whole life knowing that women can be and are represented in science. As a woman scholar myself, however, I also know that women, even when performing at least as well as their male counterparts, are often not treated the same regarding their scientific pursuits. Not being treated the same usually manifests in ways that shortchange women researchers; it often means not being taken as seriously as men or having to achieve much more to be on equal footing with male colleagues. *Meeting the Challenge* shows future generations how successful women scientists got to where they are. It offers vital glimpses into female careers and scientist lives that are not yet as common as they should be but that will hopefully become more widespread as more girls and women pursue scientific careers in environments that are more attuned to offering equitable conditions and opportunities to people of diverse backgrounds.

Any individual woman scientist will have plenty of personal experiences to recount about unequal treatment. Beyond such anecdotal evidence, there is also considerable scientific literature that shows gender inequalities in academic pursuits that exist across fields and geographic regions. Examples include receiving fewer invitations to give talks at top universities,[1] being held to higher standards when publishing in journals,[2] and being cited less even when accounting for various paper characteristics.[3] Such unjust treatment also comes through in some of the life stories featured in this book. It is a

[1] Christine L. Nittrouer, Michelle R. Hebl, Leslie Ashburn-Nardo, Rachel C. E. Trump-Steele, David M. Lane, and Virginia Valian, "Gender Disparities in Colloquium Speakers at Top Universities," *Proceedings of the National Academy of Sciences* 115, no. 1 (2018): 104–108, https://doi.org/10.1073/pnas.1708414115.

[2] David Card, Stefano DellaVigna, Patricia Funk, and Nagore Iriberri, "Are Referees and Editors in Economics Gender Neutral?," *Quarterly Journal of Economics* 135, no. 1 (2020): 269–327, https://doi.org/10.1093/qje/qjz035.

[3] Jordan D. Dworkin, Kristin A. Linn, Erin G. Teich, Perry Zurn, Russell T. Shinohara, and Danielle S. Bassett, "The Extent and Drivers of Gender Imbalance in Neuroscience Reference Lists," *Nature Neuroscience* 23, no. 8 (2020): 918–926, https://doi.org/10.1038/s41593-020-0658-y; Vincent Larivière, Chaoqun Ni, Yves Gingras, Blaise Cronin, and Cassidy R. Sugimoto, "Bibliometrics: Global Gender Disparities in Science," *Nature* 504, no. 7479 (2013): 211–213, https://doi.org/10.1038/504211a.

harsh reality that requires more work at both the institutional and the personal level for the situation to improve.

The paths girls and women have taken to successful scientific careers are varied, and this book showcases this diversity of experiences wonderfully. One noteworthy theme that emerges from the many accounts in the book is how often familial encouragement has helped girls and women forge successful paths. Another recurring theme is how support by one's male supervisors and peers can be significant. While it is important to continue working toward more just support of scientists from varied backgrounds at the institutional level, individual actors can also make crucial contributions to fostering women scientist careers.

The stories featured in this volume show that the paths women scientists take toward notable accomplishments are often winding, with unusual twists and turns rather than what a textbook success story may look like. For this reason alone, the book is significant for both aspiring scientists and those who want to support them—whether that be family members or colleagues—to know of these life stories. It is meaningful to understand that there is more than one way to pursue one's scholarly passions successfully. My mother has done a great service to the scientific enterprise more generally by showcasing the many divergent ways that women can triumph in science, especially if they are supported both institutionally and by the people around them. Among them, many examples show that having a scientific career and a family are not mutually exclusive. It is important to have such positive accounts available in a sea of scholarly literature that documents the difficulties women face in science. While it is important to be aware of discrimination and work against it, it is also imperative to offer inspiring accounts to encourage the next generation of women scientists to forge ahead.

Eszter Hargittai
PhD Sociology, Princeton University
Professor of Communication and Media Research
University of Zurich

Preface

We should not complain. Women have been represented as Science in a most beautiful manner in the most prominent of places. The figures in the following two pages show but two examples. They are from Victorian London, where science flourished and progress was all around. They are, of course, merely allegorical representations, whereas in real life there was little to boast about when it came to the recognition of women in the sciences. According to the generally accepted opinion of the time, women were good as decoration but had no place in the laboratory.

Slowly and gradually, things started to change. Today there are spectacular examples of women excelling in science and being recognized for it, too. It happened in 2020 that two women shared the Nobel Prize in Chemistry. Two men had regularly shared these prizes in the past, but this time it was two women, an unprecedented achievement. Of course, we are still early in a long journey toward equality and equal representation that may never be achieved fully.

This volume introduces successful women scientists from the past and the present, from different parts of the world and in different fields, as an expansive set of vignettes meant to locate these figures and perhaps inspire further reading. Taken together, they demonstrate that despite the obsolete attitude that "science is not for women," women can and do succeed in science, even if this success often requires courage and perseverance. This book is built on my previous volume, *Women Scientists: Reflections, Challenges, and Breaking Boundaries* (Oxford University Press, 2015), which drew primarily from my interviews with women scientists. The present volume expands the circle of heroes and brings some new and deserving names and histories to a broader readership.

Budapest, Fall 2021

"Science" by John Birnie Philip, 1868, on the façade of the Foreign and Commonwealth Office, Whitehall, London WC2. Photograph by Magdolna Hargittai.

"Science" by Farmer and Brindley, 1868-1869, on the Holborn Viaduct, London WC1. Photograph by M. Hargittai.

Acknowledgments

Many fellow scientists who figure in this book have assisted me in various ways in making this book happen. I thank them for our shared interests. All members of my family—my husband, Istvan, and our daughter, Eszter, and son, Balazs—helped me in my work. I thank the Hungarian Academy of Sciences and the Budapest University of Technology and Economics for continuous support. I appreciate the trust of my editor Jeremy Lewis, the meticulous work on the manuscript by project editor Michelle Kelley, the improvements and corrections by the copy editor Judith Hoover, and Oxford University Press (New York) for bringing out this book. I express special thanks for permission to reproduce images and apologize for any permissions to reproduce I may have improperly stated or overlooked, and if notified will make amends in later printings.

1
Astronomers

Astronomy is the oldest science. Historical records show that there were astronomical observations as early as 5000 years ago. Women became involved in astronomy very early on, often facilitated by family members who were themselves astronomers.

The first woman astronomer, whom we know as **Enheduanna**, was the priestess of the city of Ur in Mesopotamia, around 2350 BCE. She was also the chief astronomer and a poet and wielded tremendous power and prestige, as her short poem indicates:

> The true woman who possesses exceeding wisdom,
> She consults a table of lapis lazuli[1]
> She gives advice to all lands . . .
> She measures off the heavens,
> She places the measuring-cords on the earth

Alabaster disk at Pennsylvania Museum of Archaeology and Anthropology. Enheduanna is in the middle. Source: Wikipedia Creative Commons, public domain.

[1] Lapis lazuli is a semi-precious stone with an intense blue color.

"Hypatia," by Jules Maurice Gaspard and Elbert Hubbard. From *Little Journeys to the Homes of Great Teachers*, vol. 23, no. 4 (East Aurora, NY: The Roycrofters, 1908).
Source: Wikipedia, public domain.

Even that long ago there were observatories for monitoring the stars. The calendars used at the time to determine the dates of religious events were the forerunners of those we use today.

The most famous woman astronomer and mathematician from ancient times is **Hypathia** (ca. 355–415 CE). She lived in Alexandria in the Roman Empire (in today's Egypt). She had a chair at the Neoplatonic Academy, where a group of scholars discussed Platonic philosophy. She was also versed in other sciences, such as chemistry and medicine, was a celebrated teacher, and was liked and appreciated by many. She was a pagan, and her influence irritated the Christian groups. In 415, she was murdered by a mob, a volunteer militia of monks serving the archbishop. Fortunately, some of her writings have survived. The popular film *Agora* (2009) presented Hypathia, played by the British actress Rachel Weisz, capturing her contribution to astronomy.

During the Middle Ages the enterprise of science went into remission, but from around the 16th century this started to change and a growing number

Sophia Brahe.

Source: Saddhiyama; www.schoolsobservatory.org/learn/history/biographies/brahe; http://denstoredanske.dk/Dansk_litteraturs_historie/Dansk_litteraturs_historie_1/Renæssancen/Ortodoksi,_barok,_enevælde_1600-1700/De_lærde_kvinder/Ægteskab_eller_studier_-_Sophie_Brahe_og_Birgitte_Thott.)

of women became astronomers. Most of the well-known female astronomers worked together with their husbands or brothers.

Sophia Brahe

Sophia Brahe (1556–1643) was a legendary Danish astronomer who worked with her oldest brother, Tycho Brahe, making important astronomical observations. Her family disapproved of her interest because they did not find scholarly activities appropriate for the members of the aristocracy. Brother and sister were 10 years apart; Tycho wanted to teach Sophia horticulture and chemistry because he thought that astronomy would be too difficult for a woman to understand. Nonetheless, she learned astronomy on her own from German and Latin books and helped her brother with astronomical observations. Her help was essential in the discovery of a supernova (SN 1572). This work was not without risk, as the discovery added evidence against the idea that the sun, the moon, the stars, and the other planets all orbit the Earth. Another observation that Sophia made was the December 8, 1573, lunar eclipse. She and her brother were shining representatives of the Danish Renaissance.

Statue of Maria Cunitz in Świdnica, Poland.
Source: Wikipedia;: Sueroski; Zdjęcie z realnego źródła zrobione przez autora pliku.

Maria Cunitz

Maria Cunitz (1610–1664) was born into a well-to-do family in Wołow (a town in lower Silesia, southwestern Poland; then part of the Habsburg Empire). She married early, in 1623, and after her first husband died, she married Elias von Löwen, a physician, in 1630. He was a scholar of astronomy and shared his interest with her. They jointly made observations of Venus in 1627 and Jupiter in 1628.

Cunitz studied sciences, music, and history and spoke seven languages. During the Thirty Years' War (1618-1648), the couple stayed at a Cistercian convent. There, she expanded the existing astronomical tables, included all the planets known at the time, and simplified Johannes Kepler's Rudolphine Tables. After the war, she published her *Urania Propitia*, a book of astronomical tables, in German and Latin, and dedicated it to Emperor Ferdinand III. The book provided new tables and a more elegant solution to Kepler's problem concerning the determination of the position of a planet on its orbit as a function of time. Her husband emphasized in his preface to the book that it was indeed written by her and not by him.

Cunitz was well known among astronomers already in her life and received further recognition later as well. A crater on Venus is named after her, as is a minor planet, 12624 Mariacunitia. A full-size sculpture remembers her in Świdnica, Poland.

Elisabetha Hevelius

The parents of Elisabetha Hevelius (née Koopmann, 1647–1693) were well-to-do merchants in Amsterdam. They moved to Danzig (today's Gdansk, Poland), where Elisabetha was born. The young Elisabetha knew of Johannes Hevelius, a well-known astronomer, and she became fascinated with the heavens. He lived in their neighborhood, and in 1663, the 52-year-old Hevelius married the 16-year-old Elisabetha. She helped with keeping up their observatory, and he let her pursue her own interest in astronomy. Elisabetha and her husband together discovered many stars. After his death, she completed and published their joint work in a book, *Prodromus Astronomiae*—a catalog of stars—in which about 1,500 stars were listed along with their positions. Although the book was based on their joint work, it carried only his name.

Maria Margaretha Winkelmann Kirch

Maria Margaretha Kirch (née Winkelmann, 1670-1720) was a German astronomer. Her Lutheran minister father supported her interest in

Johannes Hevelius and his wife, Elisabetha, observing the sky with a brass sextant, 1673. Artist: Andreas Stech (1635–1697). Engraver: Isaak Saal. Book author: Johannes Hevelius (1611–1687). Printer: Simon Reiniger.

Source: Edited image from a digital file retrieved from the Library of Congress, https://www.loc.gov/item/78393624/; Wikimedia, public domain.

astronomy. In 1692, she married Gottfried Kirch, a famous astronomer and mathematician, 30 years her senior. They had four children, and all four learned the science of astronomy. Maria and her husband made astronomical observations jointly, and both were active in the Berlin (Prussian) Academy of Sciences. In 1700, Gottfried Kirch

Maria Kirch. Source: Wikipedia, https://thelifeofmariawinkelmann.weebly.com/introduction.html.

became astronomer royal at the Berlin court of Friedrich III. In 1702, Maria alone discovered a comet. Her husband took credit for the discovery until, about 10 years later, he admitted that *she* had discovered the comet—but it was not renamed.

Maria continued her astronomical observations and became one of the famous astronomers of her time. In 1707, she noticed the Aurora Borealis, popularly known as the Northern Lights or Polar Lights. It was another discovery when she observed the conjunction of the sun with Saturn and Venus in 1709 and the approaching conjunction of Jupiter and Saturn in 1712. A conjunction is the apparent meeting or passing of two or more celestial bodies.

In 1709, the president of the Berlin Science Academy, Gottfried von Leibniz, presented Maria Kirch to the Prussian court with high appreciation: "There is a most learned woman who could pass as rarity. Her achievement is not in literature or rhetoric but in the most profound doctrines of astronomy. . . . [I] do not believe that this woman easily finds her equal in the science in which she excels."[2]

[2] "Maria Margaretha Kirch," Wikipedia.

Her husband died in 1710, and although she was perfectly qualified to take over his work, this did not happen because there was too much resentment against employing a woman. Fortunately for her, the person who was selected to succeed her husband proved to be unqualified for the position, and, after a while, the Academy welcomed her back.

Caroline Herschel

Caroline Herschel (1750–1848), a German-born Briton, was one of the best-known early astronomers. Initially, she worked with her brother, William Herschel. When he first asked her to "sweep the sky" and find interesting objects, she was rather reluctant, but warmed to this work through habit and practice. She made her first independent discovery of a celestial body on February 26, 1783. The book of her observations is now at the Royal Astronomical Society in London.

William died in 1822, but Caroline continued her work in astronomy to the end of her life. How famous she eventually became can be seen from her many firsts. She was the first woman to hold a government position in England. She was earning a salary as a scientist at a time when even men

William Herschel and Caroline Herschel servicing their telescope. A. Diethe, color lithograph ca. 1896. Source: Wellcome Collection.

The remains of the Herschel telescope in the Garden of the Royal Observatory in Greenwich. Photograph by M. Hargittai.

were seldom paid for doing science. Her best-known achievements were her discoveries of comets, which she published between 1782 and 1787. The Royal Family was interested in her findings and invited her brother to give an account of them in Windsor Castle. Apparently, the Royals did not want to go as far as asking her to give the presentation. In 1828, she was the first woman to receive a Gold Medal from the Royal Astronomical Society, and she remained the only woman to do so for quite a while. The second woman bestowed this honor was Vera Rubin (see below) in 1996. In 1835, Herschel was named an honorary member of the Royal Astronomical Society. In this first, she shared the distinction with Mary Somerville (see below). At 96 years old, she received the Gold Medal for Science from the king of Prussia.

Mary Somerville

When Mary Somerville (née Fairfax, 1780–1872) died, the London daily newspaper, *The Morning Post*, noted in her obituary, "Whatever difficulty we might experience in the middle of the nineteenth century in choosing a king

Bust of Mary Somerville at the Royal Institution. Photograph by M. Hargittai.

Lithographic portrait of Mary Somerville, after J. Phillips. Source: Wellcome Collection.

of science, there could be no question whatever as to the *queen of science*."[3] She grew up in Scotland and had rather incidental schooling at home, in village schools, and in an academy for girls in Edinburgh. Her thirst for knowledge gradually led her to embark on independent research, in particular in mathematics and astronomy. She was a skilled writer and refined her style by studying the writings of famous scientists, such as Euclid, Newton, Laplace, and others.

She married a distant cousin in 1804 and had two children. Her husband did not think that women were clever enough to pursue scientific interests, but at least he let her study. It was a brief marriage; he died in 1807. In 1812, she married another distant cousin, Dr. William Somerville, with whom she had four children. William studied astronomy, geography, and chemistry and encouraged her to continue her research in the physical sciences. Because he had an important medical appointment, the couple moved among the best scholars and artists in London, such as the novelist and playwright Walter Scott, the early computer scientist Charles Babbage, the painter J. M. W. Turner, and, eventually, the earliest computer programmer, Ada Lovelace.

Somerville studied magnetism and light and kept publishing her findings. When she was asked to translate Laplace's monumental treatise, she did something novel. Rather than merely translating it from the French original into English, she also effectively rendered the language of algebra in plain parlance. The title was *The Mechanism of Heavens* (1830). It was an instant success and remained a textbook in Cambridge for decades. Her second book, *On the Connexion of the Physical Sciences* (1834), was an even greater success. Years later, her *Physical Geography* (1848) became the textbook of choice in English for more than half a century. Somerville continued as a great disseminator of science; her fame in this regard peaked with her fourth book, *Molecular and Microscopic Science* (1869), which would be an intriguing title even today. Her autobiography, *Personal Recollections* (1874), based on old-age reminiscences and correspondence, appeared posthumously.

Somerville, together with Caroline Herschel, were the first two female honorary members of the Royal Astronomical Society (both were enrolled in 1835). Just a few of her many posthumous recognitions: Somerville College of Oxford University is named after her; islands, squares, ships, and asteroids are named after her; a Scottish banknote bears her image.

[3] *The Morning Post* (London), December 2, 1872. See also https://www.jstor.org/stable/4025228.

Maria Mitchell

Maria Mitchell's bust by Emma F. Brigham in the Bronx Hall of Fame. Photograph by M. Hargittai.

Maria Mitchell (1818–1889) was the first American woman astronomer. She was born in Nantucket, Massachusetts, into a large Quaker family. Her parents wanted to give their children a good education. Her father was a schoolteacher and an amateur astronomer. When he noticed Maria's interest in astronomy and mathematics, he taught her how to use the astronomical instruments. He founded a school at Nantucket, where Maria was one of his students as well as his teaching assistant. When the school closed, she continued her education at the local school until she was 16 years old. In 1835, she opened her own school, introduced her own teaching methods, and allowed nonwhite children to attend her classes along with the white children. This was a brave move at the time.

In 1836, she began working at the Nantucket Athenaeum, the island's public library, where she stayed for 20 years. She could occupy herself with astronomical observations much of the time. She and her father made geographical calculations for the U.S. Coast Survey. On October 1, 1847, she discovered a comet (Comet 1847 VI). On that day, as night began to fall, she went to her telescope and started sweeping the sky. She noticed that at a place where just a moment before there was no activity, suddenly an unknown object was flying through the sky. She thought that this must be a comet, and her father agreed with her assessment. Although Maria wanted to do some calculations, he immediately wrote a letter to his friend at the Harvard Observatory about this great discovery. Due to a big storm, his letter, dated October 1, was sent only on October 3. As it turned out, even a few days back then could play a decisive role in assigning priority for this sort of discovery.

To continue the story, we have to go back a few decades, when, in 1831, King Christian VIII of Denmark decided to give a valuable gold medal to the first person who noticed a comet for which only the head of the comet could be observed. As it happened, an Italian astronomer, Francesco de Vico, discovered the same comet as Mitchell had, also in 1847, but he did so on October 3, two days after Mitchell. Although the king had died, his successor kept the promise. Fortunately, the Harvard professor, whom Mitchell's father had contacted, could see that the Mitchell letter was written on October 1, and he did everything in his power to make the Mitchell case. Vico ultimately accepted Mitchell's priority, and she received the prize from the king of Denmark. She became internationally recognized as a legitimate astronomer, and the comet became known as "Miss Mitchell's comet."

The next year she was elected as the first woman to the American Academy of Arts and Sciences, and then in the following year to the American Association for the Advancement of Science. At the time, it was uncommon to accept women into scientific societies, and on her certificate from the American Association for the Advancement of Science the words "Sir" and "FELLOW" are crossed out, replaced by "Honorary Member."

Although Mitchell did not have a college education, she was appointed professor of astronomy at what was then the Vassar Female College, becoming Vassar College in 1865, in Poughkeepsie, New York. She also became the director of the Vassar College Observatory, and stayed in this position for two decades. Her fame encouraged many women to study astronomy and mathematics.

Although she continued teaching, she never forgot her own research. With her students, she photographed the sun, trying to understand the nature of sunspots. She studied solar eclipses, nebulae, and double stars. She was a good teacher, and 25 of her students were featured in *Who's Who in America*.

Late in her life, she moved back to Massachusetts to live with one of her sisters. She passed away in 1889 and was buried in Nantucket, where there is a Maria Mitchell's Home Museum and a Maria Mitchell Observatory. Both at Vassar and in Nantucket, there are several memorials dedicated to her. She was inducted into the National Women's Hall of Fame, and a crater on the moon was named after her.

Williamina Fleming and the Women of the Harvard Observatory

Williamina Fleming (1857–1911) was born in Dundee, Scotland, and after finishing her schooling at the age of 14, she married. In 1878, she and her husband emigrated to the United States. Soon after, her husband left her and she had to support herself and her child. She became a maid at the house of Edward Charles Pickering, the director of the Harvard College Observatory. He decided to employ Fleming at the observatory rather than at his house. It

Williamina Fleming, about 1890. Source: http://www.cfa.harvard.edu/lib/online/almanac/0300c.htm; Wikipedia, public domain.

Women at the Harvard Observatory working for the astronomer Edward Charles Pickering. Author: Harvard College Observatory. Source: https://hollisarchives.lib.harvard.edu/repositories/4/archival_objects/1134143, Wikimedia, public domain.

was at this time, in 1881, that Pickering started to work on a big project, celestial photography. He put a prism in front of the lens of the telescope and captured the spectral image of the stars. Fleming's task was taking careful measurements of those images. She collected information from many thousands of stars and became famous for her work. In 1906, she became an honorary member of the Royal Astronomical Society of London, and the Astronomical Society of Mexico presented her with a medal.

Fleming was not the only woman who learned astronomy from Pickering. When he started his photography project at the Harvard Observatory, he may not have realized how much data would accumulate in his laboratory. Because of his positive experience with Fleming, he decided to employ many more women to analyze the data. Several of these women eventually became known in the astronomy community for their own work. They were the "Harvard Observatory Computers." Below I mention a few of them.

The task of **Henrietta Swan Leavitt** (1868–1921) was to examine the photographic plates and select the best ones for cataloging the brightness of the stars. She measured, for example, the brightness of the stars of the Small and Large Magellanic Clouds. She identified about 1,800 stars. **Antonia Maury** (1866–1952) published her own stellar classification catalog in 1897. It included 4,800 photographs and her analysis of 681 bright northern stars.

It was the first time that a woman was credited for an observatory publication. **Annie Jump Cannon** (1863–1941) cataloged the stars and classified them based on their spectral characteristics. Following Pickering's instructions, she excelled in organizing the information about the stars according to their spectral types and estimated temperatures.

Cecilia Helena Payne-Gaposchkin

Cecilia Helena Payne-Gaposchkin, Smithsonian Institution.
Source: Cecilia Helena Payne-Gaposchkin, Wikipedia, no known copyright restrictions.

Cecilia Helena Payne-Gaposchkin (née Payne, 1900–1979) was a British American astronomer, born in Buckinghamshire, Wendover, England. She went to a private school, and when she was 12 years old her family moved to London so her brother could get a better education. She was mostly interested in science. She received a scholarship that made it possible to continue her education at Newnham College at Cambridge University. She was still only generally interested in science when, in 1919, she attended a talk by Arthur Eddington, the famous scientist and astronomer. He told of his travels to the Gulf of Guinea, off the West Coast of Africa, where he observed and photographed the stars during a solar eclipse. Eddington's observations provided evidence for Albert Einstein's general theory of relativity. This lecture was a milestone in Payne's development. Later she referred to its impact on her as if almost suffering a nervous breakdown.

She completed her doctoral dissertation but was not granted her PhD because Cambridge did not grant degrees to women until 1948. She moved to America and became the first person to earn a PhD in astronomy from Radcliffe College of Harvard University. The accepted dogma was that the stars and the earth have the same elemental composition. In contrast, she concluded that the sun consisted predominantly of hydrogen. She was dissuaded by other experts from pursuing this conclusion, which, years later, proved to be correct.

Payne spent her entire career at Harvard University, where she had low-paid and low-ranking jobs, as women could not be appointed professor at the time. She was allowed to teach courses, but they were not recorded in the course catalog until 1945. Finally, in 1956 she was appointed professor and eventually became the chair of the Department of Astronomy—the first woman ever to serve as a department head at Harvard. When she retired from teaching in 1966 she was appointed professor emeritus. She trained a number of future outstanding astronomers. In 1934, she married a Russian-born astrophysicist, Sergei Gaposchkin, hence her double name. They had three children.

Nancy Grace Roman

Nancy Grace Roman (1925–2018) was a well-known American astronomer who worked for NASA. She was called the Mother of the Hubble Space Telescope and had a major role in developing the Goddard Space Flight Center. She was born in Tennessee; during the first years of her life her family moved frequently due to her geophysicist father's job. As a teenager, she realized that astronomy interested her most. At Swarthmore College in Pennsylvania, she learned a great deal from a Dutch astronomy professor, Peter van de Kamp. He gave her a solo lecture course on professional astronomy and guided her in learning about the instruments used in astronomy as well as how to use the library.

Nancy Grace Roman. Source: Creative Commons, NASA.

She graduated in 1946, then received her PhD in 1949 at the University of Chicago. She spent several years at the Yerkes Observatory in Williams Bay, Wisconsin, which was operated at the time by the University of Chicago. She was ahead of the Yerkes administrators in recognizing the necessity of computerization. She was a prolific researcher and published widely cited papers. One of her observations was that the stars consisting of hydrogen and helium move faster through the galaxy than the stars consisting of heavier elements. Eventually, she decided to leave the University of Chicago, where she saw no hope for professional advancement for women. In 1954, she joined the Naval Research Laboratory in Washington, D.C., and became a very active member of its fledgling radio astronomy program. Her international reputation was growing, especially when she was invited for a prestigious visit to Soviet Armenia in 1956 for the dedication of its Byurakan Observatory.

In 1959, she was appointed to head a fledgling NASA program, Observational Astronomy. With this move she became more of an administrator than a researcher, but she realized the importance of charting a project that would shape astronomy for the next half-century. She was the first woman at NASA to hold an executive position. She spent 21 years at

NASA and was a principal associate in programs that contributed decisively to the successes of the agency. In 1979, she took early retirement, learned programing, acted as a consultant in geodesic programs, taught courses, and in 1995 became head of the Astronomical Data Center at NASA's Goddard Space Flight Center.

Roman was active in the American Association of University Women. She had plenty of experience of the difficulties young ambitious women faced when they wanted to become scientists. Women who themselves were in positions such as high school counselor often did not encourage girls' ambitions, either because they felt it was not ladylike or because they wanted to protect their advisees from disappointments. Roman's successful career in science, both as a researcher and as an administrator, serves as a shining example of women opening possibilities for advancement even if this demanded extra stamina and hard work.

Vera C. Rubin

Vera Rubin (née Cooper, 1928–2016),[4] was born in Philadelphia to an immigrant Jewish family. From early on, she was interested in the stars.

Vera Rubin, 2000, at the Department of Terrestrial Magnetism of the Carnegie Institution in Washington, D.C. Photograph by M. Hargittai.

[4] Magdolna Hargittai, conversation with Vera Rubin, Department of Terrestrial Magnetism, May 16, 2000, with follow-up correspondence in May 2004. The interview was published in Balazs Hargittai and Istvan Hargittai, "Vera C. Rubin," in *Candid Science*, vol. 5: *Conversations with Famous Scientists* (London: Imperial College Press, 2005), 246–265.

Vera Rubin in the early 1970s, measuring spectra. Courtesy of Vera Rubin.

She graduated from Vassar College, where the famous astronomer Maria Mitchell became the first professor of astronomy in 1865. When Vera received her bachelor's degree, she was determined to become an astronomer. But there were not many universities at that time where a woman could get into an astronomy program. Princeton University, for example, did not allow women into its graduate program until 1975. What a contrast it was when in 2005 Princeton conferred on Rubin its honorary doctorate in a ceremony officiated by Shirley M. Tilghman, the school's first female president.

Rubin continued her education at Cornell University, where her future husband, Robert (Bob) Rubin, was a graduate student. Her master's thesis in 1951 created quite a stir. She suggested that over large distances, galaxies might be moving together. Eventually she found that the movement of the galaxies is not random; rather, they move in large clusters. It took a long time for the scientific community to appreciate this idea. Years later, her results were considered to be the first proof that the bulk of the universe is invisible dark matter.

Most astronomers rejected her idea about dark matter, but not the famous physicist George Gamow (1904–1968). He was open to her ideas and became the advisor of her PhD studies at Georgetown University. She completed her dissertation in 1954 and stayed on at Georgetown until 1965. From there, she went to work at the Department of Terrestrial Magnetism of the Carnegie Institution for Science in Washington, D.C., where she stayed for the rest of her life.

Vera Rubin's most famous work was finding evidence for the existence of dark matter. She had always been interested in uncovering the nature of galaxies, and especially the outer part of spiral galaxies, that had been little studied. It was known that near the center of a galaxy, the stars orbit at high velocity. By analogy with the planets in the solar system, it was supposed that toward the peripheries of the galaxies, the stars moved more slowly. However, Rubin and Kenneth Ford (b. 1926) together found that within galaxies, the distant stars moved as fast as the stars near the center. This was a puzzle.

Rubin's results were among the first of the recent revolutionary discoveries in astronomy. Technological advances have helped to refine astronomical observations, paralleled by advances in computation, allowing for the digestion of unprecedentedly large amounts of data. On the other hand, according to her, "there are still many major things that we don't understand. I think that every couple of years or maybe even more often we are going to learn something very, very new and important."[5] Ever since Rubin had come up with the idea of dark matter in the late 1970s, her possible Nobel Prize had become a subject of discussion among astronomers, and it still is. Her not receiving the award has been considered a serious omission on the part of the Royal Swedish Academy of Sciences. In some ways, the 2020 Nobel Prize in Physics, including the distinction of Andrea M. Ghez (see below), may be a symbolic compensation for Rubin's missing recognition. Alas, not to her, as there are no posthumous Nobel Prizes.

Back in her youth, when she was choosing a college, she was aware of the difficulties women face in scientific research. The portion of women in any given cohort decreases rapidly from the undergraduate level to higher positions. This issue was close to her heart, and she spoke about it with dedication: "You can write books just on this subject! It is perceived as a women's problem and I believe it will never be solved until it is perceived as society's problem or an academic problem. So, I am really more pessimistic now than I was 50 years ago. Then, with so many women entering college, it looked like a gradual evolution would take place. . . . In science, women now get more college degrees than men. And even in science PhD degrees, the numbers of women are now about equal to men. So I think the real problem lies in academia."[6]

[5] B. Hargittai and I. Hargittai, "Vera C. Rubin," 256.
[6] B. Hargittai and I. Hargittai, "Vera C. Rubin," 256.

She had four children. Her husband was more than merely supportive of her ambitions; he was most encouraging. When I had the opportunity to ask her about the greatest challenge in her life, she responded promptly, "Finding good care for my children." Eventually all four children became scientists. Their comments included in Rubin's autobiography in the *Annual Review of Astronomy and Astrophysics* give a hint about this extraordinary family:[7]

Dave: "One evening, when I was a child about ten years old, my mother told me that she knew something about astronomy that no one else knew. To this day, I remember thinking that this was extraordinary. . . . [W]hat my mom alone then knew was the beginning of the story of dark matter."

Judy: "We saw our parents working hard and having fun being scientists, but none of us knew at the time that we would all choose to follow their lead. . . . I feel truly blessed and deeply grateful to be able to say, 'Vera Rubin is my mother.' "

Karl: "I'm not sure when I realized that growing up in a household headed by two scientists was unusual. As a young child, I just assumed that almost all adults were scientists and that astronomy was a job for women. . . . There was never ever pressure to become a scientist, but it did seem like the natural thing to do. . . . I've learned that . . . having parents who understand and encourage such a life is an advantage most of my colleagues didn't have."

Allan: "I think it's no coincidence that the four children all ended up doing science. A pervasive early memory of mine is of my mother and father with their work spread out along the very long dining room table. . . . At some point I grew old enough to realize that if what they really wanted to do after dinner was the same thing they did all day at work, then they must have pretty good jobs."

Vera Rubin received many distinctions; only a few are mentioned here. Her memberships in learned societies included the National Academy of Sciences of the U.S.A. (1981) and the Pontifical Academy of Sciences (1996). In 1993 she received the National Medal of Science from President Bill Clinton, and in 1996 the Gold Medal of the Royal Astronomical Society (London), the second woman so distinguished. (Caroline Herschel was the first, in 1828.)

[7] Vera Rubin, "An Interesting Voyage," *Annual Review of Astronomy and Astrophysics* 49 (2011): 26–27.

My personal encounter with Vera Rubin as well as our later interactions left me with an impression of her inner harmony, which no external recognition, distinctions, or awards could have created. This is how she formulated her principal motivation: "My greatest pleasure has come from combining the roles of wife/parent/astronomer. None would have given as much joy alone. I love science because I have an unending curiosity about how the universe works, and I could not be happy living on earth and not trying to learn more. For me, it is the daily internal satisfactions that make a life in science so wonderful. Cold dark nights at a telescope have been among the greatest treasures of my life."[8]

Jocelyn Bell Burnell

Jocelyn Bell Burnell (b. 1943)[9] was born in Belfast, Northern Ireland, the first of her parents' four children. She was interested in science already in her school years. When the Soviets launched Sputnik in 1957, she got caught up in the excitement of the space race and decided to become a scientist. Why astronomy? Her father used to bring home books from the public library, but

Jocelyn Bell Burnell with the radio telescope used to detect the signals from the heretofore unknown pulsars in the 1960s. Courtesy of Jocelyn Bell Burnell.

[8] B. Hargittai and I. Hargittai, "Vera C. Rubin," 265.

[9] Magdolna Hargittai and Istvan Hargittai, "Jocelyn Bell Burnell," in *Candid Science,* vol. 4: *Conversations with Famous Physicists* (London: Imperial College Press, 2004), 638–655.

Jocelyn Bell Burnell in 2002. Photograph by M. Hargittai.

she paid only scant attention to them. However, when he brought home astronomy books, she had a different reaction: "I was hooked by the scale and grandeur and the excitement that was there in astronomy even in the late fifties, early sixties. I realized that the physics that I was learning in school was a tool that could be applied to help us to understand the cosmos."[10]

She went to school in Belfast, followed by a boarding school in England for her senior years. She attended college in Scotland and received her bachelor's degree at the University of Glasgow in 1965. Then she went to Cambridge for graduate studies in astronomy. Her supervisor, Antony Hewish (1924–2021), was just changing his research direction and decided to study radio galaxies, that is, galaxies that emit radiation in the radio-frequency range of the electromagnetic spectrum. He designed an extremely sensitive radio telescope, and his new graduate student was actively involved in its construction. It was a huge structure, built from a large number of poles and kilometers of wires. Bell Burnell was responsible for doing the measurements and recording the thousands of signals coming from the telescope. The instrument became operational in July 1967. Bell Burnell first noticed some odd signals in November of that year. This was to be the discovery of pulsars.

Such objects had never been observed, and according to Bell Burnell, the discovery was an accident. She was studying quasars; their name comes from "quasi-stellar radio sources." Describing the discovery, she used the analogy of making a video of a sunset from some splendid vantage point. Then along comes a car and parks in the foreground and has its hazard warning lights

[10] M. Hargittai and I. Hargittai, "Jocelyn Bell Burnell," 640.

going. This spoils the picture. Something similar happened in her experiment. She was focusing on some of the most distant things in the universe, and this peculiar signal popped up in the foreground. It turned out to come from a pulsar.

The peculiarity of these signals was that they were very sharp and repeated at regular intervals. At first, the team thought that there were problems with their apparatus. Then they thought that they were picking up some noise, some human-made signals from other sources. Bell Burnell even thought about the possibility that these might be messages from an alien civilization: LGM, "Little Green Men." She knew that if there ever would come a message from an outer-space civilization, it would probably be picked up by radio astronomers. Soon they found a second signal that clearly came from another location, and then more and more. They realized that the idea for the source as a faraway civilization should be dropped: the chance to receive messages from two, let alone more, civilizations at about the same time had zero probability.

Eventually, Hewish interpreted the surprising observations. The signals came from neutron stars. They have extremely high density, comparable to the density of the nucleus of the atom. We can imagine this as if the mass of the sun would be packed into a ball that has a radius of 10 kilometers (about 6 miles). The name "neutron stars" does not mean that they consist entirely of neutrons, only that they are richer in neutrons than anything else on earth. The name "pulsar" refers to "pulsating radio star." These pulsars have very strong magnetic fields and their magnetic poles do not coincide with their rotational axes, just as in the case of earth. As the pulsar spins, the beam that comes out from its magnetic pole sweeps across the sky like a lighthouse beam sweeps across the ocean. Every time the beam passes over earth, we can pick up a pulse, and this is how we get a series of regular pulses.

Soon radio astronomers all over the world were looking for pulsars and discovered many. Joseph Taylor at Harvard College Observatory was among the first to confirm the Cambridge observations. In 1974, Hewish and Martin Ryle (for an unrelated discovery in the same field) were awarded the Nobel Prize in Physics. Bell Burnell was not among the awardees, but this did not raise an eyebrow at the time—she was "just" a graduate student, after all. Still, this recognition of astronomy by this highest of awards made her happy.

About 20 years after the Nobel Prize for the pulsar discovery, something triggered people's memories and they started to talk about the unjust exclusion of Bell Burnell. In 1993, the Physics Nobel Prize was awarded for the discovery of the double pulsars. This discovery was especially important as it opened up new possibilities to study gravitation. The awardees were Joseph Taylor (b. 1941) and his graduate student, Russell Hulse (b. 1950).

There were conspicuous similarities and differences between the 1974 and 1993 Nobel Prizes. For the pulsar discovery, only the professor was distinguished, whereas for the double pulsar, the professor and his student shared the prize. The principal difference was that in the ensuing two decades, people had become increasingly aware of the contributions of younger collaborators.

Following her PhD from the University of Cambridge in 1969, Bell Burnell had quite a few positions in succession. She was at the University of Southampton for a few years, then moved to University College London, where she had a professorial appointment. She also worked at the Royal Observatory in Edinburgh. In the early 1990s, she became a professor of physics at the Open University. I met her in 2000 at Princeton University, where she was a visiting professor. In 2001, she was appointed dean of science at the University of Bath, and she stayed there for four years.

Early on, due to her husband's job, they had to move often, and this hindered her advancement. When she became pregnant, she went to the chair of the department and asked what kind of arrangements they had for maternity leave. "He said: 'Maternity leave!?! I never heard of it!'" Thus, she just resigned. For 18 years, until her son went to college, she worked part time. Being married and being a mother made a big difference for her possibilities. At the same time, she was able to get a much broader range of skills than she might have had working continuously in research.

While moving around, she managed to keep up with the developments in astronomy. She has been involved in various areas of the field, such as gamma ray, X-ray, infrared, and millimeter-wave astronomy. She acquired management training that she found very useful when she accepted the position of dean at the University of Bath.

She has received many prestigious awards and has held important administrative positions. Just a few of these: she received prizes from the American Astronomical Society, the Royal Astronomical Society, and the American Philosophical Society; she holds several honorary doctorates; in 1999, she was appointed Commander of the Order of the British Empire (CBE); in 2003, she was elected Fellow of the Royal Society (FRS, London); and in 2007, she was elevated to Dame Commander of the Order of the British Empire. Being an FRS is especially gratifying because it signals peer recognition.

In 2018, she was awarded the Special Breakthrough Prize in Fundamental Physics, worth three million dollars, for her discovery of radio pulsars. She donated the money to fund women, under-represented ethnic minority and refugee students to become physics researchers. She has become more famous by *not* having received the Nobel Prize than if she had!

France A. Cordova

France A. Cordova, 2008, in Fort Lee, New Jersey.
Photograph by M. Hargittai.

France A. Cordova (b. 1947) was born in Paris, where her Mexican American father worked for a nonprofit organization after World War II. She was the first of her parents' 12 children. Science as a profession was not yet on her mind when she went to university. As a young graduate in English, after an archaeological dig in Mexico, she wrote a short novel, *The Women of Santo Domingo*. She entered it in a contest organized by the magazine *Mademoiselle* and finished among the top 10 contestants. She then worked on the staff of the news service of the *Los Angeles Times*. At some point she decided to change fields. She earned a PhD in physics at the California Institute of Technology (Caltech), served as the head of the Department of Astronomy and Astrophysics at Pennsylvania State University at State College, and was chief scientist of NASA, the youngest person and the first woman holding this position. She was vice chancellor of the University of California at Santa Barbara and the first Hispanic chancellor of the University of California system at UC Riverside.

Following her graduation from Caltech, she spent about 10 years at the Los Alamos National Laboratory. There, she studied different aspects of astrophysics. She participated in experiments that measured the X-ray radiation from pulsars (see Bell Burnell) and from the so-called white dwarfs, the very compact, dense stars, similar to the pulsars except that their mass is smaller. She was the leader of a project to study astrophysical processes in strong gravitational fields. After Los Alamos, she moved to Pennsylvania, to her first appointment in higher education. She was moving gradually toward administrative positions.

In 2007, she became the president of Purdue University. During her time there, the College of Health and Human Sciences was established at the school. This was also when she decided to stop doing research as she felt that being the president of Purdue was a big enough challenge. Still, it was beneficial that she had a scientific background; she knew the language, she understood the problems, and could ask the right questions in planning long-range research prospects with the faculty. She was in great demand, and in 2009 she became a member of the Board of Regents of the Smithsonian Institution, where she served until 2014. This was followed by her directorship of the National Science Foundation (NSF, 2014–2020). After retirement from the NSF, she was elected to the Board of Trustees of Caltech.

We met in 2008 at a fundraiser for Purdue University. When I asked her about the tremendous shifts in her career, she said that it was not an easy path and she would not recommend it to anybody. Theirs was a large family and they always had deep discussions around the dinner table. Physics was one of the topics they liked to talk about, including the origin of the universe. Eventually, when she started to learn about the space program, she realized that this was it. Thus, she went into physics and astrophysics. Usually she was the only woman in her classes. She remembers fondly her classmates who helped her along the way.

One of her hobbies is rock climbing. During one climb, she met Christian Foster, a high school science teacher. They married in 1983. They have a daughter and a son.

France Cordova's achievements have been broadly recognized. In 1984, she was chosen by *Science Digest* to be one of "America's 100 Brightest Scientists Under 40." In 1996, during her time at NASA, she was featured in a program on PBS called *Breakthrough: The Changing Face of America*, which introduced 20 Native American, African American, and Latino scientists from different scientific fields. The next year, when she was at Penn State, she was elected to be among the "100 Most Influential Hispanics" by *Hispanics Business* magazine, and recently among the "101 Top Influential Leaders in Hispanic U.S." by *Latino Leaders* magazine. She was also among the recipients of the annual Kilby award that is given for contributions to society through science, technology, innovation, and education.

Andrea M. Ghez

Andrea M. Ghez. Source: Wikipedia.

Andrea M. Ghez (b. 1965) was an undergraduate at MIT, attended graduate school at Caltech, and became a Nobel laureate while working at the

University of California, Los Angeles—a superb set of schools and background for success. She and her associates utilized the world's best telescopes and mapped the brightest stars in the center of the Milky Way. There they observed an invisible massive object and therewith provided evidence of the huge black hole in the center.

Ghez graduated from MIT in 1987, attended graduate school at Caltech. Gerry Neugebauer (1932-2014) was her mentor, a pioneer in infrared astronomy whose father, Otto Neugebauer, was a historian of astronomy and mathematics. She received her PhD in 1992. She was awarded a Hubble postdoctoral fellowship and spent some time at the University of Arizona. She joined the University of California at Los Angeles (UCLA) in 1994, rising to full professorship in 2000. Then, in 2014, she was appointed head of the Galactic Center Orbits Initiative at UCLA. Her Nobel recognized the achievements of a 25-year-old project, which began when she joined UCLA. Her position from the start gave her the possibility of having access to the newly constructed Keck telescopes in Hawaii.

She wanted to study black holes, and of course, the immediate question is: How do you observe something you cannot see? The mass of a black hole is confined to a zero volume, hence some of the difficulties. Black holes have incredibly high density and nothing escapes their gravitational pull, not even light. She started with a small group; today her group has 30 members. Ghez and her associates provided proofs that a supermassive black hole does exist in the center of our galaxy; this means that what used to be considered a possibility has become a certainty. Still, Ghez issues a caveat that this does not signal a diminishing number of puzzles about our universe; on the contrary, the known number of unknowns is steadily increasing. David Haviland, the chair of the Nobel Committee for Physics, emphasized, "The discoveries of this year's Laureates have broken new ground in the study of compact and supermassive objects. But these exotic objects still pose many questions that beg for answers and motivate future research. Not only questions about their inner structure, but also questions about how to test our theory of gravity under the extreme conditions in the immediate vicinity of a black hole."[11]

Ghez received half of one-half of the Nobel Prize. The other half of this half went to Reinhard Genzel of the University of California at Berkeley and the Max Planck Institute for Extraterrestrial Physics. The first half of the 2020 Nobel Prize was awarded to Roger Penrose of the University of Oxford "for

[11] Andrea Ghez wins 2020 Nobel Prize in physics | UCLA.

the discovery that black hole formation is a robust prediction of the general theory of relativity."[12]

Ghez and her team have made direct measurements of gravity near a black hole. Gravity is intimately connected with Einstein's general theory of relativity. So the question arises whether the theory might provide all the necessary interpretation for including black holes or there will be a need to augment it. This is one of the questions the world of science is eager to answer. Ghez has already reported various surprises from the environment of black holes, but none has so far disputed Einstein's theory.

Prior to the Nobel Prize, one of Ghez's most prestigious recognitions was the Crafoord Prize in 2012, which, like the Nobel Prize in Physics, is awarded by the Royal Swedish Academy of Sciences. The Crafoord Prize recognizes fields that the Nobel Prizes do not cover, namely mathematics, geosciences, biosciences, and astronomy, on a rotational basis. Ghez is the fourth woman Nobel laureate in physics, joining the select group of Marie Curie, Maria Goeppert Mayer, and Donna Strickland.

The career of Andrea Ghez does not appear to have ever suffered from gender discrimination. However, the disadvantaged situation of girls in career choices must have been on her mind at least at the beginning of her career. She published a beautiful 40-page book in 1995 with the title *You Can Be a Woman Astronomer*. This book is not only for girls, but this title counters the automatic reaction that would suggest that a book on astronomy would be for boys. The book is recommended for third- to fifth-graders, though because of some specific technical terms, this may be a little optimistic, but children usually sort out for whom a book is or is not. This is a highly personal book in which Ghez speaks about her own experience, her own goals, and her own pleasures in doing astronomy. Her present fame may add a great deal to the interest in this book.

[12] The official web site of the Nobel Prize: https://www.nobelprize.org/prizes/physics/2020/summary/.

2
Mathematicians

The opportunities for women in science have been painfully slow in the making, especially when we consider the fast pace of progress in science in general. During the second half of the 19th century, however, women in Europe started to demand certain rights, and one of these was the possibility of pursuing higher education. Eventually, more and more universities opened their doors to women. The increasing involvement of women in mathematics is especially noteworthy because of the widespread view that mathematics escapes female comprehension. But the story of women and mathematics goes back much further than the 19th century.

Elena Cornaro Piscopia

Elena Cornaro Piscopia (1646–1684) was an Italian mathematician and philosopher. She is considered to be the first woman to receive a Doctor of Philosophy degree, in 1678, at the age of 32. She was born into a noble family in Venice as an illegitimate child. This meant that she was barred from noble privileges, even though her father eventually married her mother. Young Elena had a tutor who taught her languages, and by the time she was seven years old, she spoke seven languages. She was a child prodigy and was called "Oraculum Septilingue" (the seven-tongue oracle). She also studied mathematics, theology, and philosophy. Her tutor wanted her to get an advanced degree in theology, but the bishop of Padua did not allow it because of her gender. Eventually she went for a degree in philosophy. This generated a great deal of interest, and her lecture at the Cathedral of Padua attracted people not only from Padua and from the university, but even from abroad. She spoke for an hour in classical Latin, explaining Aristotle. She received her award with all the four customary insignias: a book on philosophy, a laurel wreath on her head, a ring on her finger, and a special cape over her shoulders. This was such an unprecedented occasion that the scene is illustrated in the Cornaro Window in the West Wing of the Thompson Memorial Library of

Vassar College. She was appointed lecturer of mathematics at the University of Padua in 1678 and was elected a member of academies in a number of European countries. She died in 1684 from tuberculosis. Her statue still stands at the University of Padua.

Laura Maria Caterina Bassi

Laura Maria Caterina Bassi (1711-1778) was an Italian scientist, born in Bologna. She was both a mathematician and a physicist; at the time, what is called physics today was covered by the term "natural philosophy." Bassi did not have a formal education; she had private tutors until the age of 20. She studied philosophy, logic, and natural philosophy and became interested in Isaac Newton's teachings. The archbishop of Bologna recognized her talent and became her patron. For the defense of her doctoral thesis, the archbishop arranged a public debate with four professors, and she stood the test brilliantly. She received her degree in 1732, the second woman in the world to earn a Doctor of Philosophy degree (the first was Elena Piscopia). Bassi taught at the University of Bologna and was the first woman teacher who received a salary. At one point, she was even the highest paid employee of the university.

In 1738, Bassi married Giuseppe Veratti, who was a doctor of medicine and a lecturer in anatomy at the University of Bologna. They had a large family, between eight and twelve children; the exact number is not known. Her first lecture was on water as a natural element. In 1776, she was appointed to the chair of experimental physics at the Bologna Institute of Sciences, and her husband became her teaching assistant. She passed away in 1778, at the age of 66; her health may have deteriorated due to her many pregnancies. A crater on Venus and a high school in Bologna are named after her.

Maria Gaetana Agnesi

Maria Gaetana Agnesi (1718-1799) was born in Milan into a wealthy family. She was a child prodigy. At the age of five, she spoke Italian and French, and by the age of 11 she had added five more languages: Greek, Hebrew, Latin, Spanish, and German. Her father was proud of her and did everything to further her studies. She was interested in science and studied geometry and

Bust of Maria Gaetana Agnesi, Palazzo di Brera, Milan. Source: Wikimedia.

ballistics. When she was 14 years old, her father started to organize regular meetings to which he invited the most learned men so she could discuss with them the most complex philosophical questions.

She was religious, but her father did not let her enter a convent. He stipulated that if she continued doing mathematics she could do as much charity work as she wanted. Her mother died and her father remarried twice. Eventually she had 21 siblings, and it was her task to teach them. Although this kept her busy, she always found time for mathematics. She worked with the most prominent Italian mathematician at the time, the monk Ramiro Rampinelli. She was the first woman to write a handbook on mathematics. Her most important work was the treatise "Analytical Institutions for the Use of Italian Youth."

In 1750, Pope Benedict XIV appointed her to the chair of mathematics and natural philosophy at the University of Bologna, the second woman to receive such an honor (after Laura Bassi). However, Agnesi never served; perhaps her interests changed. After her father died in 1752, she became a student of theology. She founded a home for the elderly and she herself lived there with the nuns of the institution. She died in 1799 and was buried in a mass grave for the poor.

Ada Lovelace

Ada Lovelace's memorial plaque, 12 St. James's Square, London SW1. Photograph by M. Hargittai.

Ada Lovelace (née Byron, 1815-1852) was an English mathematician, the only legitimate daughter of the famous poet Lord Byron. He wanted a son and was disappointed at getting a daughter. The marriage of Ada's parents was difficult for her mother, Anna Isabella (née Milbanke), because of Byron's infidelity. Therefore, her mother did everything possible to make sure that Ada developed interests that were as far away from Byron's as possible. Isabella herself liked mathematics (Byron liked to call her "the princess of parallelograms") and she encouraged her daughter to develop an interest in math. Mary Somerville (see among the astronomers) was one of Ada's private tutors and she and Ada became friends.

Byron left his wife and daughter and died when Ada was only eight years old. When she grew up, many of her friends were scientists, such as Michael Faraday and Charles Babbage, and writers, such as Charles Dickens. With her possibly innate interest in poetry and taste for science cultivated by her mother, she developed a unique desire to combine the two. In a letter to her mother, she wrote, "[I]f you can't give me poetry, can't you give me 'superfluous poetical science'?"[1] She developed a broad involvement in both and could use a greater imagination in mathematics than she would have otherwise. This helped when her friend Babbage came up with the idea of the programmable computer, to be used for calculations. She immediately understood its great potential.

At the age of 20, Ada married a wealthy nobleman, William, the 8th Baron King, who soon became the Earl of Lovelace, hence her better-known last name. They had three children, but this did not turn her away from mathematics. In the summer of 1833, Babbage invited her to see his "Difference Engine," a mechanical calculator designed for computing mathematical tables. She was fascinated by the machine and visited it often. Babbage called her "the enchantress of numbers." She translated an Italian mathematician's article on Babbage's new computer, titled "Analytical Engine," into English. She understood Babbage's aims and added her notes to the translation, which became three times as long as the original article.

Lovelace not only understood Babbage's goals with this new engine; she was better able to explain them than he. She could see further than Babbage had. In his world, the engines were bound by numbers, whereas she envisioned the possibility of developing and tabulating any function of any degree of generality and complexity. She recognized that the numbers could also represent entities other than quantity. This meant that if we had a machine capable of working with numbers, we could make these numbers represent things and other systems: letters or musical notes, for example. In this scenario, a number will be just one of the possible symbols that the computer will manipulate. She pronounced these views in 1843, about a hundred years before the forerunners of the present-day computer arrived.

She died in 1852, at the age of 36, due to cancer and, possibly, malpractice. The significance of her contribution to the development of the computer has been debated, but there is consensus that it has been underestimated. During

[1] Betty Alexandra Toole, "Poetical Science," *The Byron Journal* 15 (1987): 15, 55–65.

the past decades her role in the computer revolution has become better and better known. There are prestigious awards named after her; here I mention only a few examples of how her memory is kept alive. There is the computer language ADA, named after her, by the U.S. Department of Defense; there is an Ada Lovelace Award and the Lovelace Medal by the Association for Women in Computing; there is an "Ada Lovelace Day" annually and the "Ada Initiative." The *New York Times* decided to publish her obituary 166 years after her death.[2]

Sonia Kovalevsky

Sonia Kovalevsky. Source: We Are Mobile First.

Sonia Kovalevsky (1850–1891) was born Sofia Vasilevna Korvin-Krukovskaya in Moscow into a noble family of Polish German origin. She

[2] Claire Cain Miller, "Ada Lovelace: A Mathematician Who Wrote the First Computer Program," *New York Times*, March 8, 2018.

is the most famous among the first Russian female scientists. Mathematics started to interest her very early, and she was adamant in wanting to learn more of it. In her time, this was impossible for a woman in Russia, and going abroad required her father's permission, which he declined to give. Her way out was a marriage of convenience at the age of 16. She and her paleontologist husband, Vladimir Kovalevsky (1848-1935), left Russia in 1867. After her marriage her name became Sofia Kovalevskaya, but in time she changed her surname from the Russian form, indicating gender, to Kovalevsky, and her first name to Sonia.

She studied in Heidelberg, but women at that time were not allowed to graduate; the university merely let her audit lectures. Eventually she wrote three research papers on three different topics: partial differential equations, Abelian integrals, and on the rings of the planet Saturn. By then she was in Berlin, and according to her professor Karl Weierstrass, each of these papers would have sufficed for a doctorate. Years later, in 1874, she received her doctorate, *summa cum laude*, from Göttingen University. She tried but could not get a university position. She left for Sweden, where she was appointed to the honorary position of *privat-docent*, about the level of today's associate professor. Soon, however, she received an extraordinary professorship from Stockholm University—the first time for a woman in Northern Europe. (There had been two female professors in Italy.)

Sonia edited a new journal, *Acta Mathematica*, and she returned to one of her earlier interests, writing. In recognition of her activities, she received several prizes and became a corresponding member of the Imperial Russian Academy of Sciences in St. Petersburg. Still, she could not obtain a university position in Russia. She died at the age of 41 from influenza and pneumonia.

Sonia Kovalevsky was already famous during her lifetime, and when she traveled she was treated as a celebrity. The greatest mathematicians of her time considered her to be an important member of the European mathematics community. Besides this, she was regarded as a promising writer and was friends with many of the celebrated authors of her time, such as George Eliot, Anton Chekhov, and Henrik Ibsen. She was also a political activist and a fighter for women's rights. Her fame did not fade away upon her death; books have been written, films produced, stamps issued, and busts erected to memorialize her. Negative opinions have also appeared, trying to defame her with all sorts of accusations, relating to both her mathematics and her private life. Society had to get used to accepting so much from a woman—talent was supposed to be a man's privilege.

Mileva Marić-Einstein

Mileva Marić-Einstein with Albert Einstein. Source: Wikipedia.

Mileva Marić-Einstein (1875-1948) was born near Novi Sad, in northern Serbia (then it was part of the Austro-Hungarian Empire). When she was 16, her family moved to Zagreb (today, the capital of Croatia). She could enroll only as a private student at the all-male Royal Gymnasium and as a girl and private student could not be awarded a high school diploma. In 1894, she moved to Switzerland and graduated from a high school for girls there two years later. She was gifted and passed the rigorous entrance exams to enroll at what is today the Eidgenössische Technische Hochschule (Federal Institute of Technology, ETH) Zurich. There she met Albert Einstein (1879-1955), a fellow student in mathematics. Together they attended the lectures of Hermann Minkowski and other famous professors. They drew close, spending a lot of time discussing mostly mathematics. They corresponded during school holidays; a short part of one of Albert's letters to Mileva illuminates their relationship: "When I read Helmholtz for the first time, it seemed so odd that you were not at my side and today, this is not getting better. I find the work we do together very good, healing and also easier."[3] Einstein did not attend all the lectures, preferring to study at home. Marić was organized and methodological, and she helped him to focus. At the finals, the two got similar grades; at the experimental tasks she was better, while at the oral exams he was.

It is a natural question that if they worked so much together, how substantial might have been her contribution to Einstein's output? When his first paper appeared in 1900 with him as the sole author, Marić noted in a letter (according to the same source as above), "We will send a private copy to Boltzmann to see what he thinks and I hope he will answer us." It has been suggested that adding her name as an author at the time might have lowered the value of the paper.

Einstein had the strict notion that he could not marry without having a job to support his family. When Marić became pregnant, she abandoned her studies and tried, unsuccessfully, to persuade him to marry her. She returned home to her parents. Their baby girl, Lieserl, became ill with scarlet fever, leaving her scarred; she died at a very young age.

In 1902, Einstein got a job in the Bern Patent Office and in January 1903, they married. He had eight-hour workdays, but they continued doing mathematics and physics together during his free time, mostly evenings and nights, at least according to anecdotal evidence. She appeared to have anticipated his worldwide fame already at this early stage of their lives.

They had two sons: Hans Albert (born in 1904) and Eduard (born in 1910). The latter was sickly and spent a lot of time at a sanatorium. For a short while the family lived in Prague, where Einstein had a professorship at the German University. In 1912, he was appointed to a professorship at the ETH and they moved to Zurich. There was another move in 1914 as Einstein gained two positions in Berlin, though by then their marriage was falling apart, and Marić and the two boys returned to Zurich shortly thereafter. In 1916, Marić and Einstein divorced. The rest of her life she spent in virtual anonymity, whereas Einstein's was very much in the limelight. All indications suggest that she had a withdrawn and reserved character. Her fate was to be a woman who had to take care of two sons. Einstein provided

[3] Pauline Gagnon, "The Forgotten Life of Einstein's First Wife," *Scientific American*, December 19, 2018, https://blogs.scientificamerican.com/guest-blog/the-forgotten-life-of-einsteins-first-wife/

some financial support, including giving her his Nobel Prize money. It was in accordance with Marić's character that she did not produce a memoir of their time together, although there would have been interest in it and she could have benefited financially from such a publication. We are not aware of any scientific contribution by her following the time they parted. And yet, that she played a role in Einstein's early and most spectacular period remains undisputed.

Emmy Noether

Emmy Noether, before 1910. Source: Wikipedia.

Emmy Noether (1882–1935) was an outstanding German mathematician, one of the greatest of her time and, according to many, *the* greatest ever among women. She was the fourth child of the mathematician Max Noether, a professor at the University of Erlangen. She studied languages in high school and planned to become a language teacher, but soon she found her true calling in mathematics. She continued her studies in Erlangen and received her bachelor's degree in 1903. For her graduate studies, she moved to the University of Göttingen, where she attended the lectures of such giants as David Hilbert and Felix Klein. She started teaching in a low and unpaid position. According to the regulations at Göttingen University, a woman could not become a *privat-docent* (about the level of an associate professor). Klein tried to have the rules changed, but the university senate demanded in response, "What will our soldiers think when they return to the university and find that they are required to learn at the feet of a woman?"[4] Hilbert's frustration can be felt from his famous reaction: "Gentlemen, I do not see that the sex of the candidate is an argument against her admission as a privat-docent. After all the senate is not a bathhouse." For years, before she received the right to give lectures to students, her lectures were announced under the name of Professor Hilbert as "Mathematical Physics Seminar: Professor Hilbert, with the assistance of Dr. E. Noether."

Noether's most famous work is known as "Noether's theorem." It states that for every continuous symmetry in the laws of physics, there is a corresponding conservation law. Its essence is the connection between symmetries and the conservation laws. This theorem unified many ideas that had been around for some time, and mathematicians and physicists realized its fundamental importance. Einstein wrote about it with admiration. Noether received invitations to prestigious conferences worldwide.

In 1933, with Hitler and the Nazis coming to power in Germany, the Jewish Noether was fired, together with other Jewish professors. She then gave lectures to her students in her apartment. Another famous mathematician, Hermann Weyl, wrote about this time, "I have a vivid recollection of these months. Emmy Noether . . . her courage, her unconcern about her own fate, her conciliatory spirit . . . was, in the midst of all the hatred and meanness, despair and sorrow surrounding us, a moral solace."[5]

Noether accepted an invitation to be a visiting professor at Bryn Mawr College, a women's college in Pennsylvania. She enjoyed the atmosphere,

[4] Leon M. Lederman and Christopher T. Hill, *Symmetry and the Beautiful Universe* (Amherst, NY: Prometheus Books, 2004), 72.

[5] Hermann Weyl, "Emmy Noether," *Scripta Mathematica* 3 (1935): 201–210.

the teaching, and the many new friends both at the college and at nearby Princeton University. Unfortunately, this period turned out to be very short; in 1935, due to a serious illness, she passed away. Her ashes are under the walkway surrounding the cloisters of Bryn Mawr. Einstein wrote her obituary for the *New York Times*, part of which is quoted here:

> Within the past few days a distinguished mathematician, Professor Emmy Noether, formerly connected with the University of Goettingen and for the past two years at Bryn Mawr College, died in her fifty-third year. In the judgment of the most competent living mathematicians, Fraulein Noether was the most significant creative mathematical genius thus far produced since higher education of women began. In the realm of algebra, in which the most gifted mathematicians have been busy for centuries, she discovered methods which have proved of enormous importance in the development of the present-day younger generation of mathematicians. Pure mathematics is, in its way, the poetry of logical ideas. One seeks the most general ideas of operation which will bring together in simple, logical and unified form the largest possible circle of formal relationships. In this effort toward logical beauty spiritual formulae are discovered necessary for the deeper penetration into the laws of nature.[6]

Rózsa Péter

Rózsa Péter (born Rózsa Politzer, 1905–1977) was born in Budapest. Her father was a lawyer. She was interested in the sciences and started to study chemistry at the Pázmány Péter University, but soon she switched to mathematics. She attended the lectures of world-famous mathematicians, among them Leopold Fejér. Another renowned mathematician, László Kalmár, also influenced her career. In spite of her strong interest in mathematics, she had doubts about whether it was really the field she should be pursuing. She wrote, “When I began my college education, I still had many doubts about whether I was good enough for mathematics. Then a

[6] Albert Einstein, Emmy Noether obituary, *New York Times*, May 4, 1935.

Rózsa Péter. Source: Wikimedia.

colleague said the decisive words to me: it is not that I am worthy to occupy myself with mathematics, but rather that mathematics is worthy for one to occupy oneself with."[7] She graduated from high school in 1927. The Great Depression then made it hard to find a job, and her being Jewish made it next to impossible under the anti-Semitic Horthy regime. She gave private lessons and started her doctoral studies. She earned a PhD in 1932, but was not allowed to teach except in the Jewish high school. Toward the end of World War II, her life was at stake, and during the last months of the war she was incarcerated in the Budapest ghetto. After liberation, she taught at a Teachers Training College until it was closed in 1955. Then she became a professor of mathematics at Eötvös Loránd University and remained there until retirement in 1975.

[7] R. Péter, "Mathematics Is Beautiful," *The Mathematical Intelligencer* 12, no. 1 (1990): 58–64.

She was creative in mathematics. Around 1930, Kalmár suggested to her that she investigate the recent work of Kurt Gödel on incompleteness. This led to a new line of research, the study of recursive functions, in which she became an expert and which made her well known in mathematical circles. A recursive function is a function that calls itself back repeatedly and therefore involves a process of repetition until some kind of limit is reached. Today, recursive functions are widely used in mathematical logic and in computer programming. She published a number of papers that are the foundation of the theory of recursive functions. In 1932, she introduced her theory at the International Congress of Mathematicians in Zurich, with the title "Rekursive Funktionen." The well-known German mathematician Walter Felscher once noted, "In a series of articles, beginning in 1932, Péter developed various deep theorems about primitive recursive functions. . . . I admire this work, and it may well be said that she forged, with her bare hands, the theory of primitive recursive functions into existence."[8] She published the first monograph in this field, which appeared in German in 1951, *Rekursive Funktionen*. Its English translation, *Recursive Functions*, was published in 1967. Her last book was *Recursive Functions in Computer Theory*, which appeared in 1981, first in Hungarian and soon after in English. She never thought that her *Recursive Functions* would find practical applications, but it has become indispensable to the theory of computers.

She wrote books to popularize mathematics, the first during the war, *Playing with Infinity*. This book is for those, young or old, who always wanted to understand mathematics but were afraid that it was too difficult. The book has wonderful humor and has appeared, besides in the original Hungarian, in German, English, Russian, and Polish. Moreover, the Mathematical Association of America recommended the book for undergraduate mathematics libraries.

Péter received awards and recognitions for her work. Her highest accolade came in 1973, when she was elected a corresponding member of the Hungarian Academy of Sciences. Sadly, she did not live long enough to reach full membership.

[8] J. J. O'Connor and E. F. Robertson, "Rózsa Péter," MacTutor, March 2014, https://mathshistory.st-andrews.ac.uk/Biographies/Peter/#reference-12.

Kathleen Ollerenshaw

Kathleen Ollerenshaw at her home in Manchester, 2003. Photograph by M. Hargittai.

Kathleen Ollerenshaw (née Timpson, 1912–2014) lost her hearing almost completely at the age of eight. She fell in love with mathematics, graduated from Oxford University, and solved long-standing mathematical problems. She fought for establishing all-girls schools and for improving education in general. She was appointed Dame Commander of the Order of the British Empire and was elected Lord Mayor of the City of Manchester in 1975.

Her deafness was a familial condition on her father's side. She studied in a Montessori School, a small primary school, and the school arranged for her to learn lip-reading. She learned to play with numbers, doing multiplication tables, and she found this entertaining. When she turned nine years old, they got a new headmistress who was a mathematician, a graduate of Cambridge, and she paid personal attention to Kathleen's development.

At 13, Kathleen went to a girls boarding school, St. Leonards, in St. Andrews, Scotland. She was ahead of the others in mathematics but was behind in other subjects. She participated in various sports activities and joined in games that did not require hearing. At 15, she had all the credits for graduation but was told that she did not study the *proper* classes in mathematics, and in any case, there was nothing a girl could do with mathematics besides teaching. But she could not teach because she was deaf. The school suggested that she study something that would help her earn a living, but she insisted that she wanted to go to Cambridge to continue in mathematics.

She just loved the subject, but had no plans to become a mathematician: "I had no ambition to be a mathematician, I had always planned to be a husband's wife and do what my mother had done; good work, keep the house and do mathematics for fun as I always had. I had no ambition for it, but I didn't want to do anything else, I wanted to do that all the time." Her exam at Cambridge went well, but when she mentioned that she was deaf, she was told, "We won't be able to help you."[9] Next, she went to Oxford, trying to conceal that she was deaf. They asked her what she did during her summer holiday. "Now by a freak chance, the summer before, they had the League of Nations' conference in Geneva [on disarmament], and I went there with a group from our school. And by chance, they asked me to write this up for the school magazine.... Of course, that was the perfect question. I said every name correctly with authority.... [T]hey were impressed that I was a mathematician and that I was able to speak very clearly about a topic that had nothing to do with mathematics, and because of this, I got the scholarship. There only was one." She went to Somerville College in Oxford at the age of 19, receiving her bachelor's degree in mathematics three years later.

In 1936, Kathleen went to work at the Shirley Institute, connected to the cotton industry. She researched the efficiencies of different methods and ingredients used in weaving. She studied statistical techniques and used advanced algebra and was successful in problem-solving. She did not have any disadvantage due to her hearing difficulties because there was so much noise from the machines that nobody could hear anything.

Still in Oxford, she got engaged to Robert Ollerenshaw, who had been studying medicine there; they married in 1939. He soon left for his war duties as a military surgeon, and was away for three and a half years. Their son was

[9] Quotes are from Magdolna Hargittai's conversation with Kathleen Ollereshaw in 2003 in Manchester.

born in 1941. She left her job, but she missed mathematics. At the university she met Kurt Mahler, a recent German refugee mathematician, who suggested that she work on an unsolved problem. She solved it in a few days' time. Mahler persuaded her to go back to Oxford and study for a DPhil degree. She published five papers in two years, and that sufficed for her DPhil in 1945. Her daughter was born in 1946.

During the early 1950s, important things happened to her. The first hearing aids appeared, and this made an enormous difference. After living in complete silence for decades, she could hear again. The other important change was the start of her involvement in politics, especially concerning issues of education in general and the education of girls in particular. It all started when she was asked to talk at the National Council of Women. She talked about the poor conditions of schools in Manchester. Her talk caught the attention of national newspapers, and she decided to do a serious analysis of the situation of schools in the whole of England and Wales. As a mathematician, she understood that in order to be convincing enough, she had to provide numbers based on facts.

She was elected to the city council and as a member of the education committee, and then to national committees on education. She wrote widely read articles and booklets on education. During the 1960s, she toured the United States and visited schools; she wrote a series of articles about her experiences and became widely known for her work as an educator.

In 1971, her services to education were recognized: she became a dame. She was also busy in researching how women teachers who were married and had children could be attracted back to teaching after their children grew up. Around this period, tragedy struck when she learned that her daughter, Florence, had terminal cancer.

Ollerenshaw kept returning to mathematics. Later in her life, the famous Cambridge mathematician Hermann Bondi encouraged her to take an interest in "magic squares." The simplest magic square is a 4x4 array of numbers, where the sums of all rows, columns, and diagonals are the same. Her work with Bondi resulted in scientific papers and two books.[10] Later, she also became an amateur astronomer, and at the age of 79 she climbed Mauna Kea in Hawaii to observe a total eclipse of the sun.

[10] Kathleen Ollerenshaw and Hermann Bondi, *Magic Squares of Order Four* (London: Scholium Int., 1983); Kathleen Ollerenshaw, David Bree, and Hermann Bondi, *Most-Perfect Pandiagonal Magic Squares: Their Construction and Enumeration* (Southend-on-Sea: The Institute of Mathematics and Its Applications, 1998).

During her life, she occupied numerous positions, such as director, chairman, and president of various societies. She served a one-year term as Lord Mayor of Manchester. In 1978, she became the first female president of the Institute of Mathematics and Applications, succeeding Prince Charles in the position. She gave presentations on education and women's employment. She never gave up lecturing on mathematics, and in 1979 she held the prestigious Friday Evening Discourse at the Royal Institution in London on soap bubbles, the honeycomb, and other examples of beautiful symmetrical shapes. She was the second woman ever to deliver a Friday Evening Discourse. (The first was the archaeologist Joan Evans in 1923; the title of her presentation was "Jewels of the Renaissance.")

"I had a wonderful double life—mathematics and public service—and meanwhile managing Robert and our children," she told me. Although mathematics stayed with her all her life, she became much better known as an educator than a mathematician. "I don't think that there is a school anywhere in Manchester where I have not given a speech; it became as it were quite a profession." In spite of this, she stressed that even on the busiest days she never stopped doing mathematics. "Mathematics is a way of thinking. It requires no tools or instruments or laboratories. . . . Archimedes managed very well with a stretch of smooth sand and a stick for his magnificent discoveries in geometry. . . . Mathematics is the one school subject not dependent on hearing. . . . I never aspired to being a professional mathematician or to being a professional anything for that matter. If you are deaf, you are glad to 'get by.' "[11] She surely did much more than just "getting by."

Mary Winston Jackson

Mary Winston Jackson (1921–2005) was one of the women—many of them African American—who figured in the 2016 movie *Hidden Figures*. Without their heroism, NASA and its predecessor institutions could not have built up the U.S. space program. Mary Winston Jackson was born in Hampton, Virginia. She was a good student and graduated from high school with the highest honors. She continued at Hampton University, an all-Black college, where she earned bachelor's degrees in mathematics and the physical

[11] Kathleen Ollerenshaw, *To Talk of Many Things: An Autobiography* (Manchester: Manchester University Press, 2004), 229.

Mary Winston Jackson, 1979. Source: Wikimedia.

sciences. She taught at a school in Maryland and started tutoring, which she enjoyed and kept doing for much of her career. In 1944, she married a sailor in the U.S. Navy and they had two children.

In 1951, she found a job with the National Advisory Committee for Aeronautics, the predecessor of NASA. She started as a "computer" (as the people were called who worked with computers at the Langley Research Center in Hampton). About two years later, she was asked to work under an engineer in the Supersonic Pressure Tunnel. At her boss's suggestion, she enrolled in a night program at the University of Virginia, an all-white school. She had to ask permission from the city of Hampton to attend their classes. She finished the course as an aerospace engineer in 1958 and became the first Black female engineer at NASA.

Eventually she decided to leave engineering for something that she found even more important than the conquest of space, even if this meant a demotion for her. She focused on the well-being of Blacks and, generally, women, at NASA, and initiated important changes. She served on a variety of committees, helping to improve women's opportunities in science and engineering.

She received numerous awards and recognitions, among them the Apollo Group Achievement Award (1969), the National Council of Negro Women, Inc., Certificate of Recognition for Outstanding Service to the Community, several Langley awards, and the National Technical Association's Tribute Award (1976). Her latest distinction was the Congressional Gold Medal, the highest civilian award, which, sadly, came to her posthumously in 2020.

Karen K. Uhlenbeck

Karen K. Uhlenbeck in Austin, Texas, 2000. Courtesy of Karen K. Uhlenbeck.

"A woman just won the Abel Prize, known as 'math's Nobel'—for the first time," CBS News announced on March 19, 2019. Karen Uhlenbeck (née

Keskulla, b. 1942) was born in Cleveland, Ohio, to her engineer father and schoolteacher mother. As a child she loved to read, and as she grew her interests gradually moved toward mathematics. She was especially excited when she read a small paperback by George Gamow, *One, Two, Three . . . Infinity: Facts and Speculations of Science*. She studied physics at the University of Michigan, but changed to mathematics, earning her bachelor's degree in the subject. She continued at the Courant Institute of Mathematical Sciences of New York University. In 1965, she married the physicist George Uhlenbeck. At this time, the "anti-nepotism" rule was still in effect, which meant universities could not employ both husband and wife—even in different departments. George was in great demand by excellent schools, so he moved frequently, and Karen followed him. However, the places interested in him, such as MIT, Stanford, and Princeton, were not interested in hiring her. This may or may not have contributed to their marriage ending in divorce in 1976. She is now professor emeritus at the University of Texas at Austin and a visiting scholar at Princeton University and at the Institute for Advanced Study.

She has participated in outreach programs; for example, she cofounded the Women and Mathematics Program at the Institute for Advanced Study, "with the mission to recruit and retain more women in mathematics." She has been elected a member of the National Academy of Sciences of the U.S.A. (1986). The following quote reveals the motivation for her National Medal of Science (2000): "Uhlenbeck made pioneering contributions to global analysis and gauge theory that resulted in advances in mathematical physics and the theory of partial differential equations. She is considered a founder of geometry-based analytical methods. She is also a leader in encouraging young women to study mathematics."[12] The pinnacle of her recognition in mathematics was the Abel Prize, which she received "for her pioneering achievements in geometric partial differential equations, gauge theory, and integrable systems, and for the fundamental impact of her work on analysis, geometry, and mathematical physics."[13]

[12] https://www.scientificamerican.com/article/national-medals-of-scienc-2000-11-15/.

[13] https://www.princeton.edu/news/2019/03/19/uhlenbeck-receives-abel-prize-geometric-analysis.

Maryam Mirzakhani

Maryam Mirzakhani. Image credit: Stanford News Service.

Maryam Mirzakhani (1977–2017) was an exceptionally talented Iranian mathematician who passed away at the age of 40 from breast cancer. She was born in Tehran. Her talent was recognized early, and she won the gold medal for mathematics at the Iranian National Olympiad when she was still in high school. During subsequent years she won several more medals, one at the International Mathematical Olympiad in Hong Kong (1994) and another in Toronto (1995).

After receiving her bachelor's degree from Sharif University of Technology in Tehran, she continued her studies in the United States. She was awarded her PhD in 2004 from Harvard University. She remained in the United States and had appointments at the Clay Mathematics Institute in Providence, Rhode Island, at Princeton University, and, finally, as a professor of mathematics at Stanford University. In 2008, she married a Czech computer scientist, Jan Vondrak; they had one daughter.

In 2005, at the age of 28, the Brilliant Minds Foundation acknowledged her as one of the top 10 young minds who have tried to find innovative directions. One of her principal interests was hyperbolic space, which is different from the usual Euclidean space. According to Euclid, one, and only one, line parallel to a given line can pass through a fixed point, whereas in non-Euclidean hyperbolic space, there is no limit of parallel lines that can pass through such a point.

During her short life, Mirzakhani received a large number of awards and other recognitions. She won the Ruth Lyttle Satter Prize in Mathematics for women who make outstanding contributions to the field. She became a foreign associate of the French Academy of Sciences (2015), was elected to the American Philosophical Society (2015), and became a member of the National Academy of Sciences of the U.S.A. (2016) and the American Academy of Arts and Sciences (2017). She received the prestigious Fields Medal in 2014. It is awarded to mathematicians under the age of 40 by the International Mathematical Union every four years at the International Congress of Mathematicians. Mirzakhani tragically passed a few short years later.

3
Physicists

The Loneliness of the Female Physicist

We open this chapter about women in physics with a few photographs displaying gatherings of physicists. It may take a few moments to spot the sole female participant. (In one case, there are two.) Many similar group photographs could be presented with exclusively male participants. Our purpose is to demonstrate the loneliness of female physicists. Recalling such a pronounced imbalance gives us a glimpse of the courage and determination of women scientists who paved the way to a more equitable world today. Some information about Yvette Cauchois and Hilde Levi is given following the image in which they appear; others are introduced further below.

Marie Curie (front row, third from left) among the participants of the 1927 Solvay meeting of theoretical physicists. Back row: A. Picard, É. Henriot, P. Ehrenfest, Éd. Hersen, Th. De Donder, E. Schrödinger, E. Verschaffelt, W. Pauli, W. Heisenberg, R. H. Fowler, L. Brillouin. Middle row: P. Debye, M. Knudsen, W. H. Bragg, H. A. Kramers, P. A. M. Dirac, A. H. Compton, L. de Broglie, M. Born, N. Bohr. Front row: I. Langmuir, M. Planck, Curie, H. A. Lorentz, A. Einstein, P. Langevin, Ch.-E. Guye, C. T. R. Wilson, O. W. Richardson. Photograph by Benjamin Couprie. Courtesy of Dominique Bogaerts and the International Solvay Institutes, Brussels.

Lise Meitner is in the front row at a Copenhagen physics conference in 1937. Left to right: Niels Bohr, Werner Heisenberg, Wolfgang Pauli, Otto Stern, Meitner, Rudolf Ladenburg, unknown. Another woman physicist, the German Jewish refugee scientist Hilde Levi, also attended this meeting. She is seen at the top of the image, at the extreme left in the last row. Courtesy of the Niels Bohr Archive, Copenhagen.

The participants—50 male physicists and Maria Goeppert Mayer (middle, second row)—of the fifth Washington Conference on Theoretical Physics at George Washington University in January 1939. It was during this meeting that Niels Bohr informed a group of 24 physicists about the discovery of nuclear fission. Goeppert Mayer was the sole female physicist in the room. George Washington University, Astrophysics Group. Courtesy of Wendy Teller.

Yvette Cauchois (front row, third from left) among the participants of the ninth Solvay Conference on Physics, 1951, Brussels. Front row: N. P. Allen, Cauchois, C. O. G. Borelius, W. L. Bragg, C. Møller, F. Seitz, J. H. Hollomon, F. C. Frank. Middle row: C. W. Rathenau, W. Köster, E. Rudberg, L. Flamanche, O. Goche, L. Groven, E. Orowan, W. G. Burgers, W. Shockley, A. Guinier, C. S. Smith, U. Dehlinger, J. Laval, E. Henriot. Back row: R. Gaspart, W. M. Lomer, A. H. Cottrell, G. A. Holmes, H. Curien. Photograph by G. Coopmans. Courtesy of Dominique Bogaerts and the International Solvay Institutes, Brussels.

The sole woman physicist in her group shown in the figure is **Yvette Cauchois** (1908–1999). She was born in Paris, studied at the Sorbonne, and received her degrees in physics, including her doctorate in 1933. She worked with the Nobel laureate Jean Baptiste Perrin (1870–1940) and was his successor as head of the Physical Chemistry Laboratory. She founded the Centre de Chimie Physique at the University of Paris at Orsay. Her principal research interests were in X-ray spectroscopy, X-ray optics, and various applications of the synchrotron.

Hilde Levi is in the fourth row, fourth from left, at the Commemoration Meeting for Niels Bohr in Copenhagen, July 9–13, 1963, among a large number of physics luminaries. Source: Wikipedia.

Hilde Levi (1909–2003), the only woman in her group in the figure, was a graduate of the University of Munich. She earned her doctorate at the Kaiser Wilhelm Institute for Physical Chemistry and Electrochemistry in Berlin. She had to flee Germany because of the anti-Jewish measures imposed by the Nazis. She spent the next decade at Niels Bohr's Institute of Theoretical Physics in Copenhagen. In 1943, when Germany occupied Denmark, she had to flee again, and until the end of World War II she worked in Stockholm. After the war she spent the rest of her career in Denmark. She learned the radiocarbon technique of dating in the United States and used it in her research. She made important contributions to the history of science and wrote George de Hevesy's biography.

The Radium Institute

The Radium Institute in Vienna—the Stefan Meyer Institute for Subatomic Physics of the Austrian Academy of Sciences. Photograph by M. Hargittai.

The discovery of radioactivity at the end of the 19th century greatly increased the importance of the Joachimsthal mines in northern Austria (today, in the Czech Republic); the mines became the world's largest supplier of radium and uranium ores. The Institute for Radium Research (Institut für Radiumforschung, widely known as the "Radium Institute") in Vienna opened in 1910 with Franz Serafin Exner (1849–1926) as director and Stefan Meyer (1872–1949) as acting director. In 1920, Meyer became the director; he stayed in this position until the *Anschluss* (the annexation of Austria by Nazi Germany) in 1938. The Institute became one of the three significant research centers on radioactivity in the world. The other two were the Curies' Radium Institute in France and Rutherford's laboratory at Cambridge University. Due to the close proximity of the Joachimsthal mines, Vienna's Radium Institute was in a privileged position, helping it to foster connections with the other research venues. Many famous scientists came to visit and collaborate with its associates.

A large number of female scientists worked at the institute between 1910 and 1938; of the approximately 160 scientists, 66 (35 percent) were women. Some of them occupied important administrative positions. There was a pleasant atmosphere, and the gender of the researchers did not play a role in their advancement. The associates of the Institute were interested in each other's projects and results and assisted one another as needed.

Friendships continued beyond the Institute. The personalities of Meyer and Associate Director Karl Przibram contributed much to creating this atmosphere. Four of the notable women researchers were Elisabeth Rona, Marietta Blau, Elisabeth Karamichailova, and Berta Karlik.

Elisabeth Rona

Elizabeth Rona (1890–1981) was born in Budapest. Her physician father was the first doctor in the country to use X-ray machines in medical practice as a diagnostic tool. She wanted to become a medical doctor, but while her father encouraged her interest in science, he thought that being a physician would be too difficult for a woman. Instead, she studied chemistry, geochemistry, and physics at the University of Budapest and received her PhD in 1912.

For postdoctoral training, she went abroad and visited a variety of schools. She worked with the Polish (later Polish American) chemist Kasimir Fajans (1887–1975) at Karlsruhe University. Fajans was one of the pioneers of the science of radioactivity. When World War I broke out, Rona returned to Hungary and worked with the future Nobel laureate George de Hevesy (1885–1966), who introduced the radioactive tracing method to follow

Elizabeth Rona in the laboratory. Courtesy of the Hans Pettersson Archive-Gothenburg University Library.

chemical and biochemical events. Rona worked with isotopes and is credited with coining the terms “isotope labels” and “tracers.”

There were turbulent times in Hungary following World War I: first the revolutions, then the White Terror, the poisoning consequences of the Trianon Peace Treaty, and the flaring up of vicious anti-Semitism. De Hevesy had to flee, and the Jewish Rona adopted a somewhat peripatetic lifestyle during the early 1920s. She first moved to Berlin and worked alongside Otto Hahn at the Kaiser Wilhelm Institute for Chemistry. Then she moved to the Institute for Fiber Chemistry. In the mid-1920s she arrived in Vienna, where she became an associate of the Radium Institute and an expert in nuclear science. Together with Berta Karlik she analyzed the half-lives of radioactive elements, investigated radioactive decay, and contributed to the development of radioactive dating. For these studies, Rona and Karlik were awarded the Haitinger Prize in 1933 by the Austrian Academy of Sciences. This prize is awarded for studies in chemistry and physics that have practical use for industry. This fruitful period ended with Germany’s annexing Austria in 1938. The Jewish members of the Radium Institute were dismissed immediately, among them, Stefan Meyer. Following brief periods in Stockholm, Oslo, and Budapest, in 1941 Rona fled to America.

She was first a visitor and then, in 1948, became a naturalized U.S. citizen. Following a brief period of unemployment, she received a teaching post at Trinity College in Washington, D.C. Thanks to a Carnegie Fellowship, she could engage again in research and utilize her special knowledge and experience. She analyzed seawater, river water, and sediments, and determined their radon content. Her skills in nuclear chemistry were based on experience in the best European laboratories and working alongside such famous scientists as Fajans in Karlsruhe, Otto Hahn in Berlin, de Hevesy in Budapest, Meyer in Vienna, and Irène Joliot-Curie in Paris. Gradually her opportunities broadened and she was a most welcome addition to the Manhattan Project. Being an immigrant, though not yet a citizen, she could not be an official member of the team; however, she was involved in solving problems without knowing about the bigger picture. She was happy to contribute to the war effort of her adopted country. When the U.S. government wanted to buy her method of polonium extraction, she gave it without compensation to the Office of Scientific Research and Development.

Another area of her experience was the possible consequences of human exposure to radiation. In the early days, scientists working with radioactivity were not quite aware of its dangers. Rona was an exception. While working at the Radium Institute with radioactive materials, she requested protective gas masks. When this was denied because it was generally believed that this work carried no danger, she purchased the protective masks with her own money. She wrote extensively about the damages the human body suffers if exposed to radiation. Her methods of investigation were used in the studies conducted at the Office of Human Radiation Experiments—a rather infamous and dark side of nuclear research in the United States.

After the war, Rona returned to Washington to teach at Trinity College. In 1947, she joined the Argonne National Laboratory near Chicago. She carried out studies in nuclear science and published her findings for the U.S. Atomic Energy Commission. When she became a U.S. citizen, she was appointed senior scientist at the Oak Ridge Institute of Nuclear Studies in Tennessee. She retired in 1965 at the age of 75 and took up teaching again, this time at the University of Miami, until her second retirement, in 1976. She then returned to Tennessee and worked on her book about nuclear science and her experience with it.[1] Some delayed acknowledgment was accorded to her when she was inducted into the Tennessee Women's Hall of Fame in 2015 and the *New York Times* published her obituary in 2019.[2]

Marietta Blau

Marietta Blau. Courtesy of Walter Kutschera. Copyright Eva Connors.

Marietta Blau (1894–1979) was born in Vienna into a well-to-do Jewish family. She studied at the University of Vienna and received her PhD in 1919 for her work on the absorption of gamma rays. Franz Serafin Exner was her advisor; he introduced research in radioactivity to Austria. Blau spent a few years at an industrial firm in Berlin and in the Institute of Medical Physics

[1] Elizabeth Rona, *How It Came About: Radioactivity, Nuclear Physics, Atomic Energy* (Oak Ridge, TN: Oak Ridge Associated Universities, 1978).

[2] Veronique Greenwood, "Overlooked No More: Elizabeth Rona, Pioneering Scientist amid Dangers of War," *New York Times*, August 28, 2019, updated January 3, 2020.

at the University of Frankfurt. In 1923, she returned to Vienna and started working at the Radium Institute.

In 1925, Blau worked out the photographic method for detecting nuclear events. Cloud chambers had previously been used in such research, but her photographic method was better suited to recording rare nuclear processes due to its integrating effect. In 1932, she invited her former student **Hertha Wambacher** (1903–1950) to work with her. Together, they continuously improved the photographic emulsions in their experiments. Their first important result was to show the presence of recoil protons from the recently discovered neutrons. Eventually they turned their attention to cosmic rays. Austrian physicist Victor Hess (Nobel Prize, 1936), the discoverer of cosmic rays, introduced them to the research station of Hafelekar, near Innsbruck, 2,300 meters above sea level. Blau and Wambacher's new photographic plates were left at the station for four months before they were developed. The high-altitude location minimized interference of the cosmic rays with other objects before reaching the photographic plates. Blau and Wambacher analyzed the tracks of cosmic rays on the photographic plates and assigned them the role of indicating "disintegration stars." They published their discovery in *Nature* and wrote a longer report for a more specialized journal. For this work, in 1937 Blau and Wambacher received the Ignaz L. Lieben Prize of the Austrian Academy of Sciences.

Blau understood that she did not have a future in the Institute. This had not only to do with the increasingly anti-Semitic atmosphere in Austria but also within the Institute itself. Several of the physicists were members of the Nazi Party or sympathized with the Nazis. When Blau and Wambacher were preparing their paper for publication, one of their colleagues told Blau that she should put Wambacher, an early member of the Nazi Party, as the first author. Blau did not comply and left Vienna in 1938, just days before the *Anschluss*. In the following years, she worked in Norway, Mexico, and the United States. In 1954, at the age of 60, she finally returned to Vienna. Today there is a plaque in front of the Radium Institute honoring her and other famous scientists who worked there. Another plaque honoring her is on the wall of her former high school.

In 1950, the Nobel Prize in Physics was awarded to the British physicist Cecil Powell "for the development of the photographic method of studying nuclear processes and his discoveries regarding mesons made with this method." Glaringly, the first part refers to the discovery that Blau and Wambacher made. Powell himself wrote in his autobiography that he started using photographic emulsions for detecting particles *only after* reading Blau and Wambacher's paper about this method.[3] Wambacher died in 1950, but Blau was still alive, and according to the Nobel Archives, she received five nominations between 1950 and 1957. She could have been included with Powell, who received the prize alone. Apparently there were disagreements within the committee concerning her inclusion. There are different ideas why she was not included in the prize; one is that the discovery of the pi meson was the major reason for Powell's prize. However, reading what is available about the nominations, the impression forms that she was not treated fairly, and that racial and gender biases surely played a role in the Nobel decision.

Elisaveta Karamichailova

Elisaveta Karamichailova. Source: Wikimedia.

Elisaveta Karamichailova (1897-1968) was a Bulgarian physicist born in Vienna, as both her parents studied there—her father, medicine, and her

[3] Cecil Powell, "Fragments of Autobiography," University of Bristol, 1987, 16, https://www.bristol.ac.uk/physics/media/histories/12-powell.pdf.

mother, music. In 1909, the family moved to Sofia, Bulgaria. She graduated from an all-girls school in Sofia and moved back to Vienna for her university studies. In 1922, she received her PhD in physics and mathematics. She joined the Radium Institute, where her main research interest was radio luminescence. She studied the transmutation of light elements. Transmutation happens when one element changes into another under, for example, nuclear radiation.

In 1931, she and Marietta Blau observed a specific type of previously unknown radiation emitted from polonium. British physicist James Chadwick confirmed it as neutron radiation, eventually leading to his discovery of the neutron, for which he received the 1935 Physics Nobel Prize. When Karamichailova's paid position at the Radium Institute expired, she stayed for two more years, unpaid. In 1935, Girton College of Cambridge awarded her a three-year Alfred Yarrow Research Fellowship, which brought her to the famous Cavendish Laboratory. In 1939, she moved back to Bulgaria, where she became an associate professor at Sofia University. For a while, lacking the necessary equipment during the latter years of World War II, she could not continue her research. Conditions improved after the war, and she eventually earned the title of professor, continuing her work on radioactivity first at Sofia University and later at the Bulgarian Academy of Sciences.

Berta Karlik

Berta Karlik (1904–1990) was an Austrian physicist. She came from an upper-class family, was taught at home, and besides fulfilling the school curriculum, she learned foreign languages. She studied at the University of Vienna and became associated with the Radium Institute. Her specialty was the use of the scintillation counter, an instrument to measure radiation. She received her PhD in 1928 and thereafter was appointed to a visiting position at the laboratory of W. Lawrence Bragg in England, where she learned crystallography. Upon her return to Vienna in 1931, she continued working at the Radium Institute. She worked with Hans Pettersson, the Swedish physicist and oceanographer. They tried to determine the biological effects of uranium contamination in seawater. The non-Jewish Karlik was among the few who did not have to flee the Radium Institute after the Nazi takeover of Austria. She and another woman scientist, **Traude Bernert** (1915–1998), discovered three isotopes of the short-lived element No. 85, which is known

Berta Karlik memorial by Thomas Baumann in the arcade courtyard of the University of Vienna. Photograph by M. Hargittai.

today as astatine. After World War II, Karlik continued her work at the Radium Institute and was appointed its director. In 1950 she received the title of associate professor and in 1956, professor of experimental nuclear physics. In 1973, she was elected to the Austrian Academy of Sciences, its first female full member.

Marie Curie

Marie Curie (née Sklodowska, 1867–1934) is the best-known female scientist, and not only in the scientific literature but in popular literature, too. She was born in Warsaw and moved to Paris at the age of 24 to continue her studies. She met the physicist Pierre Curie and they fell in love. Initially, she declined his proposal to get married because she planned to return to Poland to teach, but eventually they married and lived in France.

Pierre and Marie Curie around 1903. Source: Wikipedia.

Marie Curie in the 1920s. Source: Wikipedia.

People often wonder which of the two was the greater scientist. This is impossible to know. He was eight years older and already a respected scientist with several discoveries behind him, especially on magnetism and crystals. Still, he remained relatively unknown, was not yet a member of the French Academy, and did not publish much. Most of what we know of his early works is from Marie Curie's writings. She started to work on radioactivity—a name she coined—and he joined her when he realized its importance.

Marie and Pierre Curie wonderfully complemented each other. He was a quiet, withdrawn person and she was vivacious. He never showed outbursts of enthusiasm over an idea, and the only time he acted quickly was when he proposed to her. He liked to work with family—first with his brother, Jacques, and later with his wife. They became the perfect partners, both in their private life and in science. She had the ability to bring out the successful scientist in him. She was a fast thinker, not afraid of making bold conclusions and declaring them. She wanted to make sure that their work was getting proper recognition. From the beginning, they carefully delineated in their publications who did what, and she published under her name alone as well. When the question of the Nobel Prize came up in 1903, it was impossible to ignore her. Pierre and Marie Curie shared half of the physics prize "in recognition of the extraordinary services they have rendered by their joint researches on the radiation phenomena discovered by Professor Henri Becquerel," who received the other half of the prize. In his Nobel lecture, Pierre specifically pointed out her results by saying "Mme. Curie showed," "Mme. Curie has studied," and the like. When publishing a joint paper, he wrote, "Mme. Curie and I." Marie did not deliver a Nobel lecture on this occasion, but she did get a second opportunity, in 1911, when she received an unshared Nobel Prize in Chemistry for a number of discoveries related to chemistry and radioactivity. In her lecture she paid tribute to Pierre, beautifully and movingly. She remains the only scientist to receive two Nobel Prizes in two different science categories.

Right after World War I broke out, Marie, together with her 17-year-old daughter, Irène, organized X-ray machines for hospitals and for the front to help in locating shrapnel and other metal fragments in soldiers' body. Initially, nobody saw the importance of this diagnostic work, but they continued to work as X-ray technicians for two years, when, finally, even the government recognized the importance of using X-ray machines. By the end of the war, their machines had examined more than one million soldiers and saved a great many of them.

From early on, Marie Curie was interested in applications of radioactivity to medicine. They called it "radiotherapy" (in France, *curiethérapie*), used mostly against cancer. Marie Curie died in 1934 from aplastic anemia. This is a condition in which the body does not produce enough new blood cells. Besides this illness, others, like leukemia, might result from radiation exposure. When radioactivity was first discovered, nobody thought of the possibility of its adverse health effects. The Curies carried vials of radium and other radioactive materials in their lab coats and kept them in their desks. They enjoyed looking at them and admired their bluish-greenish glow. Only gradually, due to unexplained illnesses and the deaths of young colleagues with similar symptoms, did people start to worry. It took a long time to recognize the problem and even longer to make people understand the imminent and long-term dangers.

Marie Curie and her colleagues absorbed huge amounts of radiation. Her personal effects are still radioactive and will remain so for another 1,500 years! She is buried in the Panthéon in Paris, together with her husband. It is the burial place of the greats of France, like Westminster Abbey in London and the Novodeviche Cemetery in Moscow. She was the second woman buried in the Panthéon and was the first buried there on her own merit. Her radioactive body rests in a coffin lined with an inch-thick layer of lead to absorb the radiation.

Isabelle Stone

Isabelle Stone, 1920. Source: Wikipedia.

Isabelle Stone (1868–1944) was an American physicist and one of the founders in 1899 of the American Physical Society. Of the 40 founding members, two were women, Stone and Marcia Keith of Mount Holyoke.

Stone was born in Chicago, and education was important in her family. She got her bachelor's degree in 1890 at Wellesley College, a women's school in Wellesley, Massachusetts. Upon returning to Chicago, she worked as a physics teacher and started doing research for her doctoral degree at the University of Chicago. She investigated the physical properties of thin metallic films prepared in a vacuum. She received her PhD in 1897 after defending her thesis, "On the Electrical Resistance of Thin Films." She taught at several schools, and between 1908 and 1914 she and her sister ran a school in Rome for American girls. Between 1915 and 1923, she was head of the Physics Department at Sweet Briar College, a private school for women in Virginia. In 1900, she attended the first International Congress of Physics in Paris, one of only two women among 836 participants. The other was Marie Curie.

Harriet Brooks

Harriet Brooks. Source: Creative Commons.

Harriet Brooks (1876–1933) was the first Canadian nuclear physicist. She had eight siblings, but only she and one of her sisters attended college.

The family moved often before settling in Montreal. Brooks enrolled at McGill University in 1894, just a few years after women were first allowed to attend. She graduated in 1898 with a first-class honors degree in mathematics and philosophy and received a special prize for her outstanding performance in the former.

She was fortunate to be at McGill when Ernest Rutherford arrived, and she became his first graduate student. The topic of her master's work was electricity and magnetism, and its results were published in 1899 in the *Transactions of the Canadian Section of the Royal Society*. In the same year, she was appointed a nonresident tutor at the Royal Victoria College, a new women's college at McGill. In 1901, she was the first woman to receive a master's degree from the university. She continued working with Rutherford on the radioactive emissions from the element thorium.

For her doctoral studies, Brooks attended Bryn Mawr College in Pennsylvania. This liberal arts school was founded in 1885, the first women's college offering a PhD program. There, she received a prestigious Bryn Mawr European Fellowship, and Rutherford arranged for her to go to the Cavendish Laboratory at the University of Cambridge as its first female fellow. Upon her return to the Royal Victoria College, she continued to work with Rutherford until she took up an appointment to the faculty of Barnard College in New York, another women's college.

When she got engaged to a Columbia University professor, the dean of Barnard, a woman, stipulated that the marriage should terminate Brooks's official relationship with Barnard College. There followed a heated exchange of letters, and the dean, backed by the trustees of the college, won. According to the trustees, a married woman professor could not be a successful academic. Brooks valued her professorship so much that she broke off the engagement and stayed at Barnard.

In 1906, she traveled to Europe, met Marie Curie, and joined the Radium Institute in Vienna. Brooks later married Frank Pitcher, a physics instructor; they settled back in Montreal, and this time she ceased her involvement with science.

Harriet Brooks contributed significantly to the early development of nuclear science. She was the first person to show that the radioactive substance emitted from thorium was a gas. She also did pioneering research on radon and actinium. In 2002, she was inducted into the Canadian Science and Engineering Hall of Fame.

Lise Meitner

Lise Meitner, 1946, lecturing at the Catholic University in Washington, D.C. Source: Wikipedia.

Lise Meitner (1878–1968) was born in Vienna into a middle-class Jewish family with eight children and grew up in a stimulating intellectual atmosphere. She wanted to pursue higher education, but women at that time were excluded from the university. Even studying for the *Matura* (a higher-level secondary school graduation, required for entry into university) was out of reach. Fortunately, the rules changed near the turn of the century. She passed all the necessary exams and finally, in 1901, enrolled at the University of Vienna. She was interested in mathematics and physics, and after stepping into a physics laboratory, she knew that this was her calling. She became especially interested in experimental work. She received her doctorate in 1906 (the second woman in Austria to earn a doctorate in physics) and started to look for a job, the prospects of which were not promising. She ultimately accepted a teaching job in a girls' school, and during the evenings she worked with Stefan Meyer at the Radium Institute, learning experimental procedures in the young field of radioactivity.

Meitner realized that there were no opportunities for her as a researcher in Austria. At the age of 28, she still depended on the support of her parents; fed up, she moved to Berlin in 1907. For the next 30 years she was remarkably fortunate to work at the Kaiser Wilhelm Institute, a rare example of a woman earning for herself in academia. For a while she had to work in the basement because women were not allowed to work in the laboratories, and there was no women's toilet either.

From early on, Meitner collaborated with Otto Hahn, a chemistry professor at the Institute. This partnership was advantageous for both of them, with their different backgrounds, hers in physics and his in chemistry. Both became respected members of the international community of nuclear scientists. In 1917, she was asked to establish a new physics department in the Institute. From then on, she and Hahn worked separately, although they had a number of joint projects. She was appointed full professor in 1925.

Medal by Ernst Nordin honoring Lise Meitner, 1999, commissioned by the Royal Swedish Academy of Sciences. Photograph by and courtesy of Balazs Hargittai.

The first decades of the 20th century witnessed the beginning and thriving of the new nuclear science. It started with the discovery of radioactivity by Henri Becquerel in 1896 and was followed by a large number of discoveries and several Nobel Prizes. In 1934, Enrico Fermi and his associates at the University of Rome bombarded uranium, the heaviest natural element, with neutrons. They concluded that the products of the bombardment contained two new elements that were heavier than uranium, called "transuranium elements." Meitner and Hahn, just as other scientists in the field, worked on the transuranes for about four years and published their results extensively.

In the meantime, dark clouds gathered over Europe. In March 1938, Nazi Germany annexed Austria and thus Meitner became a German citizen and, being a Jew, was forced to flee. She received an invitation from Manne Siegbahn to join his new nuclear physics institute in Stockholm. Meanwhile, Hahn and his young colleague, the analytical chemist Fritz Strassmann, continued the experiments the three of them had started, and tried to establish what elements were produced by hitting uranium with neutrons. Hahn updated Meitner about the new experiments through correspondence; he badly needed her expertise for interpreting the results.

In December 1938, Fermi received the Nobel Prize in Physics for, among other accomplishments, his demonstration of the existence of new radioactive elements, heavier than uranium, produced by neutron bombardment. At about the same time, Hahn and Strassmann found that such transuranium elements were not formed in the experiment. Rather, they observed lighter elements as a result of neutron bombardment of uranium. This was shocking news, and Hahn asked Meitner to interpret this new observation. According to Meitner and her refugee physicist nephew, Otto Robert Frisch, fission of uranium produced two lighter elements. Their rudimentary calculations supported their idea of nuclear fission. Having this reassurance, Hahn and Strassmann published their findings. They did not include Meitner among the authors, nor did they give credit to Meitner and Frisch as the first ones to recognize that nuclear fission had taken place.

In Copenhagen, Frisch provided additional experimental evidence in support of the idea of nuclear fission. Meitner and Frisch published a joint paper about it, complemented by a paper by Frisch alone about his experiments. All these milestone papers appeared at the very beginning of 1939. Hahn started insisting that he and his chemistry alone made the pivotal discovery, implying that they no longer needed any help from physics, thus belittling Meitner's contribution. The discovery of nuclear fission opened the way to the utilization of nuclear energy, for the production of both weapons and peaceful energy.

The first atomic bombs expedited the conclusion of World War II, tragically so for two Japanese cities in 1945. In that same year, Hahn was awarded the Nobel Prize in Chemistry for his discovery. (The 1944 award was postponed because of the war.) The members of the nuclear science community knew about Meitner's crucial role in this discovery, and discussions of Meitner's missing Nobel recognition have never stopped.

Meitner stayed in Sweden doing research until 1960. Her working conditions were less than ideal, but fleeing to Sweden in 1938 saved her life. She was elected a foreign member of the Royal Swedish Academy of Sciences

(RSAS) in 1945, and became a full member in 1951 after obtaining Swedish citizenship. She was the second woman elected to the Academy. In 1999, the RSAS issued a medal honoring Meitner, the first time that a female scientist had been so distinguished.

Lise Meitner memorial by Thomas Baumann in the arcade courtyard of the University of Vienna. Photograph by M. Hargittai.

Statue of Lise Meitner by Anna Franziska Schwarzbach, 2014, in the Court of Honor at Humboldt University, Berlin. Photograph by M. Hargittai.

Meitner received many international awards during her life and posthumously, far too many to list here. In 1957, she received the peace class of Pour le Merité, the highest German award for a scientist. Many national academies of sciences elected her a member, among them the Austrian Academy of Sciences in 1966, as the first female member of its science class. She is remembered in sculptures and plaques in both Austria and Germany. The most distinguished posthumous recognition for her is that the element of atomic number 109 is now called Meitnerium, Mt, a more unique distinction than even the Nobel Prize.

Leona Woods Marshall

The American physicist Leona H. Woods—later, Leona Woods Marshall Libby—(1919-1986) was the only woman on the team which developed the world's first sustainable nuclear chain reaction.[4] She was a versatile scientist whose contributions enriched nuclear physics, astrophysics, archaeology, and environmental protection. She was born on an Illinois farm in La

Leona Woods Marshall, 1946. Courtesy of U.S. Department of Energy.

[4] When a neutron bombardment splits a large atomic nucleus, uranium, for example, into two smaller ones (this is fission), a large amount of energy is liberated simultaneously. During this process statistically more than one neutron may emerge for each neutron used in the bombardment, and this leads to a nuclear chain reaction. Above a certain "critical mass" of the fissionable material, a certain isotope of uranium, for example, the nuclear chain reaction develops into an explosion—this is how atomic bombs are produced. Below such a "critical mass," a sustainable nuclear chain reaction happens, which is the basis for energy production in atomic power stations.

Leona Woods Marshall with her colleagues, participants in creating the world's first nuclear reactor. Front row, from left: Enrico Fermi, Walter Zinn, Albert Wattenberg, Herbert Anderson. Middle row: Harold Agnew, William Strum, Harold Lichtenberger, Leona Woods Marshall, Leo Szilard. Top row: Norman Hilberry, Samuel Allison, Thomas Brill, Robert Nobles, Warren Nyer, Marvin Wilkening. Courtesy of U.S. Department of Energy.

Grande, not very far west of downtown Chicago. Her father was a lawyer and her mother a homemaker, and Leona was the second of their five children. She attended the local Lyons Township High School, which lists her among its notable alumni. She graduated from high school at the age of 14, enrolled at the University of Chicago, and graduated from it with a bachelor's degree at the age of 18. She stayed at the University of Chicago for graduate studies and became the disciple of the future Nobel laureate Robert S. Mulliken. She did both experimental and high-level theoretical work in diatomic molecular spectroscopy and was awarded her PhD in 1943 when she was only 23 years old.

Beginning in 1942 she participated in the nuclear project, utilizing her exceptional talent for experimental work in developing the world's first nuclear reactor in Chicago. Her specific task was developing boron trifluoride counters for detecting neutrons in the chain reaction once the reactor became operational. Fermi charged her with recording his lectures for the crew of physicists and engineers building the reactor. Fermi was an excellent lecturer, and this was a great educational experience for her.

The reactor was constructed under the stands of the abandoned football stadium, Stagg Field, in its squash court where Woods used to play squash. When the reactor went operational on December 2, 1942, she was there, among some 20 participants and observers. It was a historic moment; many count this the beginning of the atomic age. It was also a solemn moment, as all those present recognized that what happened was a prelude to what may yet come as they were liberating a heretofore unknown and untamed source of energy. There was also a feeling of anxiety and apprehension that if they could do this, so could the Germans—a terrifying thought.

There were no speeches, no official celebrations. The physicist Eugene P. Wigner produced a fine bottle of Chianti along with some paper cups and poured a taste for each one present. As if attesting to their participation, all signed the straw-roped Chianti bottle. This became the only authentic record of the day's event. Some of the participants gathered in 1946 to remember that day. In 1962 there was another collection of signatures. This remembrance took place on November 27 in Washington, D.C., and again on December 1 at the University of Chicago. Leona Marshall tells of this in her excellent book, *The Uranium People*.

In 1943, she married a fellow physicist, John Marshall. They had two boys. Leona and John separated soon thereafter, and she effectively became a single mother, despite the fact that they officially divorced only in 1966. Even during her pregnancies, she continued working on the nuclear project, perhaps not always observing what we would today recognize as adequate safety precautions; the full extent of the hazards was not known at that time.

After World War II, Leona returned to the University of Chicago as an associate member of Fermi's new Institute for Nuclear Studies. She was appointed assistant professor in 1953. Four years later, she moved to the Institute for Advanced Study in Princeton, and then, in 1958 to the Brookhaven National Laboratory in Upton, New York. Gradually her interest was shifting from nuclear processes to elementary particles. In 1960 she became assistant professor of physics at New York University and was appointed professor two years later. Her next move was to the University of Colorado, where her interests broadened to include astrophysics. She joined the RAND Corporation, a research and development think tank established to assist the American armed forces. She stayed with RAND until 1976.

Following her divorce from John Marshall, she married the physical chemist and 1960 chemistry Nobel laureate Willard Libby (1908–1980). Libby was a pioneer of radiocarbon dating, an efficient tool of archaeology. At the time of their marriage, he was a professor at the University of California, Los Angeles. From this time, she called herself Leona Woods

Marshall and eventually joined her husband at UCLA as a visiting professor. By now her interests included a host of engineering fields, archaeology, environmental protection, and food safety. She became a prolific author. In addition to 200 or so papers, she published seven books between 1969 and 1983. Her topics included the creation of an atmosphere for the moon, issues of environmental protection, the story of some aspects of the Manhattan Project (*The Uranium People*), cosmology, and Willard Libby's life and work.

Maria Goeppert Mayer

Maria Goeppert Mayer with a slide rule. Courtesy of U.S. Department of Energy.

Maria Goeppert Mayer (1906–1972) was the second female Nobel laureate in physics. Comparing the numbers of women receiving Nobel Prizes in the science categories, physics fares the worst. During the 120 years of the Prize, altogether four women have received it: Marie Curie in 1903, Goeppert Mayer in 1963, Donna Strickland in 2018, and Andrea Ghez in 2020.

Maria Göppert was born in Kattowitz (then, Germany; today, Katowice, Poland). When she was four years old, the family moved to the famous German university town Göttingen, where her father, Friedrich Göppert, was appointed professor of pediatrics. She was an only child and her father taught her not to grow up to be "just a woman"; she should find interests outside the typical female roles of wife and mother. Early on, she was attracted to mathematics and the sciences. Göttingen was one of the best places in the world for these fields at the time, with a large number of world-famous scientists at the university.

When Maria was a university student, Max Born, a leading physicist, invited her to attend his course on quantum physics. It was then a revolutionary branch of science, and Born was one of its principal architects. Friedrich Göppert died in 1927, and the memory of her father further reinforced her drive to become a university professor. She wrote her doctoral thesis under Born's mentorship.

Maria married the American visiting scientist Joseph E. (Joe) Mayer in 1930, and from then on was known as Maria Goeppert Mayer. The newlyweds moved to the United States, where the chemist Mayer had been offered a position at Johns Hopkins University in Baltimore. They had two children, Maria Ann and Peter Conrad. Goeppert Mayer did not let family duties slow down her research; indeed, she did not become a housewife.

This was a difficult time for many women in science in the United States due to the anti-nepotism rule, which prevented Maria from getting a job at the same school where her husband worked. It did not matter that her knowledge of quantum mechanics was superior to that of all the physics professors at that time at Johns Hopkins. Still, she had some influence at the university, and she did not need many resources as her interest was theoretical physics. She participated in the scientific activities of the school and collaborated with several faculty members, including Mayer. Her connection to Alfred Sklar was especially successful. Goeppert Mayer and Sklar published important papers in which she applied her knowledge in quantum mechanics to problems in chemistry. She and her husband together produced a book on statistical mechanics, which became a classic of the field.[5]

[5] Joseph Mayer and Maria Goeppert Mayer, *Statistical Mechanics* (John Wiley and Sons, 1940).

During the first years of their American life she returned to Göttingen several times and continued her joint work with Born and others. When the Nazis came to power, the Jewish Born and most other world-renowned scientists became refugees; her visits stopped. Goeppert Mayer and one of her colleagues at Johns Hopkins helped refugees in their efforts to relocate. In 1937, Mayer did not get tenure at Johns Hopkins and the family moved to New York, where he was hired by Columbia University. She still could not get a job except for an unpaid lectureship, even though she was at the forefront of modern physics and often the only woman at gatherings of physicists.

In January 1939, she attended the fifth Washington Conference on Theoretical Physics at George Washington University in Washington, D.C. (see image earlier). Each of these annual meetings focused on a single topic; that year, it was low-temperature physics and superconductivity. At this meeting, Goeppert Mayer was the only female physicist among her peers, including some of the greatest scientists of the 20th century. The highlight of this meeting was Niels Bohr's announcement of the discovery of nuclear fission in Berlin (see more about it under Lise Meitner). Today, a memorial tablet commemorates the historical event at George Washington University.

An important moment for Goeppert Mayer's scientific development was the arrival of Fermi at Columbia University. Her involvement in nuclear physics intensified due to his presence. When World War II reached the United States, it opened up new opportunities for female scientists. Sarah Lawrence College, a small women's school in New York State, offered Goeppert Mayer a teaching position, and the Nobel laureate Harold Urey invited her to participate in a war-related secret project. Her participation eventually became a full-time job, so she had to resign from teaching. This was the first time she was paid for her research. In 1944, Edward Teller involved Goeppert Mayer in a secret project for the Manhattan Project at Los Alamos, though she would perform this work in New York City. Teller asked her to investigate radiation transfer through various materials. He could not tell her the purpose of their study, but when he told her the temperature for which she had to carry out the calculations, she understood the implications.

After the war, in 1946, Mayer was offered a position at the University of Chicago and the family moved there. Goeppert Mayer had a professorship there, still unpaid, but at least it was a respectable position. When the Argonne National Laboratory was established, its director, Robert Sachs—Goeppert Mayer's first (informal) doctoral student back at Johns Hopkins—offered her a position of senior physicist on a half-time basis. Finally, she had a peacetime research job for which she was paid.

Meanwhile, Teller had turned his attention to the origin of the elements and invited Goeppert Mayer to work with him. This work required a serious mathematical background and she was the right person for it. Their cooperation led her to the discovery of the shell structure of the atomic nuclei, for which she received a share of the 1963 Nobel Prize in Physics. Her most successful model of atomic nuclei involved magic numbers and an onion-like shell structure, for which some labeled her "the Madonna of the Onion." Just about the same time she described the model, a research group in Germany, led by Hans Jensen, published a similar model, and hence the two would share half of the Nobel Prize "for their discoveries concerning nuclear shell structure."[6]

A few years earlier, in 1960, the Mayers received an offer from the University of California in San Diego to hire both of them (though with only half-pay for her). By this time, she was so famous that, after all these years, the University of Chicago had finally offered her a full-time position with full salary. The Mayers nonetheless decided to move to California because the climate attracted her. She already had serious health concerns. A few years before, she had suffered a stroke, leaving her partially paralyzed. Even after the stroke, she continued to work until the last days of her life.

Goeppert Mayer's career reflected the difficulties and the slowly changing opportunities for women scientists during the middle of the 20th century. She did not receive any awards or recognitions for her work before her Nobel Prize. In 1965, she was elected a fellow of the American Academy of Arts and Sciences and received the Golden Plate Award of the American Academy of Achievement. It is only too human that Goeppert Mayer felt bitter about having to work for free for most of her now-storied scientific life. In a rare candid moment, she opened up about herself to a young woman just starting her scientific career and warned her about how hard it would be to find a job in the same venue as her husband, and how her own marriage had given many in academia an excuse not to treat her seriously. Maria Goeppert Mayer had to excel not only in physics but also in turning at least some of the disadvantages of being a woman into advantages.

[6] Eugene P. Wigner received the other half for uncovering the role of symmetry in nuclear physics.

Antonina F. Prikhotko

Antonina Prikhotko and her husband, Aleksandr Leipunsky, 1928, in Leningrad. Courtesy of Boris Gorobets.

Antonina Fedorovna Prikhotko (1906–1995) was well known among physicists in the Soviet Union, but hardly at all outside the world of scientists. She never traveled abroad, did not even try; she found it humiliating to go through the process of obtaining permission for it. She declined every invitation to join the Communist Party to protect her privacy. She was the only female physicist full member of the Ukrainian Academy of Sciences and, however politically incorrect this may sound, she was beautiful.

She was born into a Cossack family. In 1923, she enrolled at the Leningrad[7] Institute of Technology, majoring in physics. There she met a fellow student, Aleksandr Ilyich Leipunsky, and they married in 1926. He came from a

[7] Today it is St. Petersburg, as it was prior to Soviet times as well.

Jewish family, and that, in time, brought them additional burdens on top of all the other difficulties in Soviet life. In 1930, the couple moved to Kharkov (Kharkiv, then, the capital of Ukraine), where they became associates of the Ukrainian Physical-Technical Institute (UFTI). Leipunsky was appointed director of the Institute, but terrible times were coming. The period 1936–1938 has become known as Stalin's Great Terror, when many people—often those in high military, party, or governmental positions—were arrested on trumped up charges, sentenced in show trials, and executed right away or exiled to slave-labor camps for many years. Some of the brightest physicists of UFTI met such a fate, and Leipunsky was anticipating that his turn might also come. He was a leading physicist on a national scale, and under his directorship the UFTI had become a leading institution in physics in the Soviet Union. His visibility certainly added to his being a candidate among the "enemies of the people." He and Prikhotko discussed what they should do in case he were arrested. In anticipation of that and other unknowns, they sent their daughter to Prikhotko's relatives, far from Kharkov.

One day in 1938, the secret police did arrest Leipunsky. The following day, Prikhotko publicly condemned her husband for having stopped being vigilant in directing UFTI and letting German agents penetrate the institute. She categorically disassociated herself from him. Leipunsky might have anticipated the harshest sentence for his "crimes." However, his case happened during the subsiding period of the Great Terror, and after two months of incarceration, he was let go. We can only suspect that Prikhotko's behavior upon her husband's arrest was choreographed for such an eventuality. Upon his return, they continued their harmonious married life as if nothing had happened. When their daughter returned, she did not notice anything different, and learned only years later about what happened in her absence. People did not freely discuss such matters in those days.

Prikhotko had an excellent mentor in her student days, Ivan V. Obreimov, a future full member of the Soviet Academy of Sciences who encouraged her to embark on independent research. She chose the investigation of cryocrystals, substances that solidify only under extremely low-temperature conditions. Solid oxygen was her favorite substance of inquiry. She soon acquired her PhD-equivalent scientific degree, and in 1943 she became the first female doctor of science (DSc) in the physical-mathematical sciences in the Soviet Union. The DSc degree has no strict equivalent in American academia, and the habilitation of the German system is not very close to it either. The DSc degree in the Soviet Union, and now in Russia, corresponds to significant research achievements and is a condition for a professorial appointment at a university or to be a laboratory head at a research institute. Prikhotko's feat was all the more significant because she completed and defended her DSc dissertation while she and her laboratory were under evacuation during the war years. This was in the city of Ufa, 1,436 kilometers (892 miles) northeast of Kharkov. She and her colleagues worked on projects helping defense efforts. Prikhotko brought with her to Ufa the experimental material on which she based her DSc dissertation.

Upon her return from evacuation in 1944, Prikhotko organized a division of crystal physics in the Institute of Physics in Kiev, which had in the meantime become the capital of Ukraine (today, Kyiv). She continued doing pioneering scientific work and is credited with the experimental discovery of molecular excitons. This is the movement of excitation in crystal structures from one crystal cell to another. Subsequently, other physicists worked out the theory of molecular excitons on the basis of her experimental discovery. In 1966, the work on excitons earned a group of physicists, among them Prikhotko, the Lenin Prize. It was the highest distinction, which could be received only once, unlike other prizes.

She was not only an outstanding physicist; she was also a friendly human being. She was respected, and her unyielding demeanor in her interactions earned her the label of being an "absolutely iron lady." This, however, did not reduce her popularity among her colleagues, and when the Institute of Physics was undergoing a crisis, the president of the Ukrainian Academy of Sciences invited her to serve as the Institute's director. The leadership of the Academy understood that only a bona fide scientist would have sufficient authority to ensure that all coworkers followed directions. She accepted the challenge for a five-year period, between 1965 and 1970. She promised only what she could deliver, and then she delivered what she promised. For quite a while, after she was no longer director, she remained the principal scientist of the Institute and, for many, a role model.

Chien-Shiung Wu

Chien-Shiung Wu (1912–1997) was born in Liuhe, Taicang, in Jiangsu province, China. Her father advocated gender equality and founded one of the first schools in China that admitted girls. He instilled the value of education in his daughter. In 1934, Wu received her bachelor's degree in physics

Chien-Shiung Wu, 1963. Courtesy of the Smithsonian Institution Archives.

U.S. postage stamp honoring Chien-Shiung Wu, 2021.

from the National Central University in Nanjing, graduating at the top of her class. She worked in research for a few years in China, then moved to the United States and earned her PhD in 1940 from the University of California, Berkeley. In 1942, she married Luke Chia-Liu Yuan, the grandson of the first president of the Republic of China.

During World War II most American physicists worked on defense-related projects, so there was a shortage of physics instructors. In 1943, Wu was appointed to a teaching position at Princeton University. Her appointment was remarkable considering that at that time women were not allowed to study at Princeton. Her teaching time there did not last long, however, because the following year she joined the Manhattan Project to work on radiation detectors at Columbia University. After the war, Wu continued at Columbia's Physics Department in nuclear physics for the rest of her career. Her main interest was beta-decay, one of the so-called weak interactions associated with radioactive decay.

She is remembered most of all for her role in the discovery of the so-called parity violation. Parity is a fundamental symmetry property; in everyday terms, it refers to the relationship of a particle or a process to its mirror image. Our right hand mirrors our left; a right-handed screw and a left-handed screw are mirrored. Similarly, a particle spinning clockwise is the mirror image of a particle spinning anticlockwise.

Two Chinese-born American theoretical physicists, T. D. Lee and C. N. Yang, wrote a paper in 1956, "Question of Parity Conservation in Weak Interactions," discussing the possibility of parity violation and suggesting experiments to test it. Wu was an expert in one of the suggested experiments. She contacted colleagues at the National Bureau of Standards (NBS) in Washington, D.C., who had the necessary equipment to carry out this experiment, and a fruitful collaboration developed. They did observe parity violation, and when it happened for the first time at the NBS, Wu rushed to Washington to see it with her own eyes. Soon after, two other experiments independent of the one at the NBS provided additional evidence for parity violation.

In the fall of 1957, Lee and Yang were awarded the Nobel Prize in Physics for anticipating parity violation. Many people think that since Wu's team was the first to provide experimental evidence for parity violation, she should have been among the awardees.[8] She is highly respected among physicists,

[8] In one category up to three people may share the Nobel Prize, so, there was an "empty slot" in the Physics Nobel Prize of 1957.

called "the Chinese Madam Curie" and "the First Lady of Physics." Her Nobel omission might indicate another case of gender bias. However, a closer examination[9] of what happened revealed a more complex story and reduced suspicion of bias in the Prize. She made an outstanding contribution to bringing down the axiom of parity conservation in weak interactions; however, according to the rules of the Nobel Prize, the awarded work must have been published before the year of the prize, in this case before January 1, 1957. Therefore, *none* of the experimentalists could have been considered for the prize that year. Of course, the award committee could have waited a year, but with the three experiments and the large number of physicists involved, it would not have been an easy choice. Thus, the physics committee may have chosen the easiest way and honored the two theoreticians who first suggested (although did not prove, of course) the possibility of parity violation in their 1956 paper.

There are plenty of cases in the history of science when talented women were truly denied the opportunity to do research, to participate in university life, or to receive proper recognition for their achievements. Wu was not one of them. She was a remarkable scientist, and with her perseverance, her thirst for knowledge, her experimental skills and rigor, and her dedication to her students, she was, and will remain, a wonderful role model for all young people aspiring to a career in physics. From early on, she was highly respected; by the end of her career, she had received an extraordinary number of prizes and other distinctions. To list a few of the most prestigious:

Fellow, American Physical Society (1948)
Member, National Academy of Sciences of the U.S.A. (1958)
National Medal of Science (1975)
First female president of the American Physical Society (1975)
First person selected for the Wolf Prize of Physics (Israel, 1978)
National Women's Hall of Fame (1998)

Ruby Payne-Scott and Other Australian Female Physicists

Ruby Payne-Scott (1912–1981) was born in Grafton, New South Wales, Australia. She attended high school in Sydney and graduated with honors in

[9] Magdolna Hargittai, "Credit Where Credit's Due?," *Physics World*, September 2012, 38–42.

Ruby Payne-Scott. Source: Creative Commons.

mathematics and botany. She studied mathematics, physics, chemistry, and botany at the University of Sydney, receiving her BSc in 1933. She was the third woman to graduate in physics in Australia. She worked at the Cancer Research Laboratory of the university, investigating the effects of the earth's magnetism on living organisms.

At the beginning of World War II, most physicists were busy with the war. This is why Payne-Scott could find a job at the Commonwealth Scientific and Industrial Research Organization (CSIRO). She was performing top-secret research on radar technology as part of the Radiophysics Laboratory and became an expert in distinguishing Japanese aircraft signals from other sources. This helped in tracking the planes from farther away, even at night. The radar was widely used in Europe, but it did not work quite as well in the southern hemisphere. Payne-Scott figured out that the reason was the tropical weather, so she created a device that improved radar detection of aircraft.

After the war, she was much involved in repurposing radar technology for peaceful ends, including scientific research. She focused her attention on the radio bursts originating from the sun and constructed new devices to facilitate the observation of these signals. In 1951, she married and ended her scientific career in order to start a family, there being no maternity leave in those days. However, she did attend the 1952 conference of the International Union of Radio Science at the University of Sydney as the sole female participant. In 2018, the *New York Times* published her obituary in its series "Overlooked No More."[10]

When Payne-Scott worked at the CSIRO, there were three highly qualified women physicists at the agency. The other two were **Rachel Makinson** and **Joan Maie Freeman.** During World War II, all three worked with radar technology. At the end of the war, their work changed. Payne-Scott remained with radar and then radio astronomy. Makinson worked on wool in the Division of Textile Physics of CSIRO. There was not much information about the properties of wool fibers and about its processing. It became Makinson's life work to study the physics of wool fibers and how they interact at the microscopic level. She was a world expert and the first woman to become a chief research scientist at CSIRO.

Freeman received a senior studentship after the war, which made it possible for her to study for her PhD at the University of Cambridge. She was the first female recipient of the Ernest Rutherford Medal and Prize in 1976. She passed away in 1998. Posthumously, in 1999, she was appointed Officer of the Order of Australia "for service to science in the field of nuclear physics and to the environment as an advocate for social responsibility in scientific research."

Rosalyn Yalow

The banquet following the award ceremony is another of the Nobel Prize festivities. In 1977, Rosalyn Yalow (née Sussman, 1920–2011) was one of the three winners in the Physiology or Medicine category. According to tradition, one of the laureates in each category gives a two-minute speech toward the end of the dinner. On this occasion, Yalow gave this speech in her category. Usually, when the time comes for the talk, a student goes to the dinner

[10] Rebecca Halleck, "Overlooked No More: Ruby Payne-Scott, Who Explored Space with Radio Waves," *New York Times*, August 29, 2018, https://www.nytimes.com/2018/08/29/obituaries/ruby-payne-scott-overlooked.html.

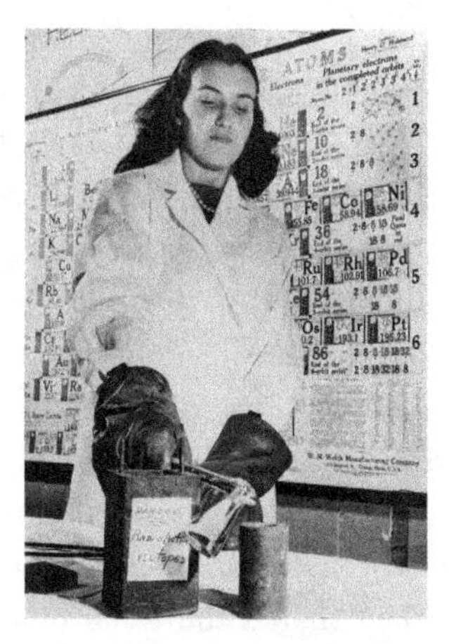

The young Rosalyn Yalow in the laboratory. Courtesy of Rosalyn Yalow.

table and escorts the laureate to the podium. That day, the student checked the seating chart, saw that there were two Yalows, and without hesitation he went to Aaron Yalow's place, opposite the King of Sweden and Rosalyn Yalow. Rosalyn understood the situation, smiled, and walked unescorted to the podium. The student realized his mistake and caught up with Rosalyn at the end of the table. The student's mistake was understandable; a woman Nobel Prize winner was still a rarity.

Rosalyn Yalow was born in New York City into a poor Jewish family. Her parents came from Eastern Europe and lacked even a high school education, but they were determined that their children go to college. Rosalyn was an outstanding student, and she decided to become a scientist. After high school, she entered Hunter College, then the women's branch of the City University of New York, which at the time charged no tuition. Her favorite subject was physics. She wanted to go to graduate school, but that was not easy without funding. While still at Hunter, she landed a secretarial position at Columbia University with a bonus: Columbia let its employees attend classes tuition free. In 1941, she graduated *summa cum laude* as Hunter's first physics major. The war-time circumstances increased her chances of getting into graduate school because so many men were enlisted. She went to the University of Illinois at Urbana-Champaign. Among all the faculty plus teaching assistants at the College of Engineering, she was the only woman. One of the students in the entering group of physicists was Aaron Yalow, her future husband.

They married in 1943, while still in graduate school. Aaron did not have research ambitions, became a college professor, and always supported her career goals. As a physics PhD, Rosalyn accepted a job offer from the Bronx Veterans Administration (VA) Hospital. Her task was to set up a radioisotope service. A janitor's closet served as her first office. She designed and built equipment for radiation detection.

Rosalyn Yalow recognized the enormous potential in applying radioactivity in medicine, well beyond the context of cancer treatment. But she needed a physician partner, skilled in internal medicine. She found him in Solomon A. Berson (1918–1972), a new resident doctor at the VA Hospital. Thus began their 22 years of joint work, which led to discoveries that changed endocrinology forever. They shared the same office and were discussing things all the time. Benson taught Yalow everything that she needed in biology and medicine, and she taught him some physics. Compared with a university setting, they had the great advantage of having no competition between each other, and no other competitors around.

She was rather aggressive, determined, and stubborn; Berson was also a rather aggressive person, but they tolerated one another. She saw signs in him of a sense of male superiority, and she let him take the spotlight, which came naturally to him because he was a leader in whatever he did. Berson was also fair to her in what really mattered. He made sure that credit and recognition were given where they were due. They alternated first authorship of their papers; they went to conferences together, and both participated in the discussions. He considered her for the most part an equal partner; still, he was also old-fashioned, and at social events he suggested that she sit with the wives.

Although they planned experiments together, she set them up and carried them out. She did much of what might be considered secretarial work: she made the plane reservations and did the necessary paperwork. In their scientific research, they wonderfully complemented each other: he was broad-minded and romantic; she was logical, mathematical, precise, and practical. The first important work they did together was the development of what would become known as the Yalow-Berson method for the in vivo determination of human blood volume. They introduced a radioactive tracer[11] into the bloodstream and by monitoring its decay rate could determine trace

[11] A chemical compound, in whose molecules at least one of the atoms is replaced by its radioisotope.

amounts of components in the blood. This was crucial for investigating hormones, because they are present in only very small amounts.

Their techniques were so revolutionary that they had difficulty getting their manuscripts accepted by the scientific journals. Leading journals rejected them, and a less prestigious journal accepted them only after they agreed to soften their claims. Their most important discovery was the determination of insulin in the blood using what they called radioimmunoassay (RIA). An immunoassay is a biochemical test that measures the presence of a macromolecule through the use of an antibody. In the case of an RIA, molecules labeled by radioactivity are used. RIA is an unbelievably sensitive method and is also inexpensive. The discovery was a breakthrough in biomedicine. When they discovered RIA they decided not to patent it; on the contrary, they did everything to help it spread, even organizing courses to teach physicians to use the technique.

Yalow and Berson could have stopped there, but they understood that they had a method in their hands whose applicability was practically limitless. They continued their work with the same energy as before and found numerous other applications, such as measuring vitamins, steroids, prostaglandins, tumor antigens, enzymes, and viruses in the blood. They also measured the hepatitis B virus, and with this work they brought RIA into the fight against infectious diseases. By measuring human growth hormones, they determined whether or not children needed to get growth hormone treatments at an early age.

At the end of the 1960s, Berson decided to move on; a few years later, in 1972, he died of a heart attack. This was a tragedy for Yalow. Many outsiders and even some of their close colleagues thought that he was the brains behind all their discoveries; they did not think that Yalow would be able to continue on her own. There had been talk about a possible Nobel Prize for their team; this no longer seemed feasible. There had never been a Nobel Prize given for a surviving partner of a team, and there is no posthumous Nobel.

Yalow was stubborn and determined to prove herself one more time. She named their laboratory after Berson so her publications would still carry his name. She took over many of his engagements and continued their research with her usual dedication. During the next four years she and her new young coworker, Eugene Straus, published about 60 papers and made several discoveries. In 1975, she was elected to the National Academy of Sciences of the U.S.A., and in 1976 she received the prestigious Lasker Award.

Rosalyn Yalow and Carl XVI Gustaf, King of Sweden, on December 10, 1977, at the Nobel banquet. Courtesy of Rosalyn Yalow.

In October of the following year, the call finally came from Stockholm: she was awarded half of the Nobel Prize in Physiology or Medicine for the RIA of peptide hormones.[12] Yalow was the second woman to receive the Physiology or Medicine prize, after Gerty Cori in 1947, and the first American-born woman to receive a science Nobel. A large number of honorary degrees and awards followed, including the National Medal of Science 11 years later.

In the mid-1990s, Yalow's health deteriorated, but still she remained active, continuing to give talks about the importance of science and on issues related to women in science. She advocated the establishment of daycare centers at universities so that young mothers would not need to be away from their research for too long.

Rosalyn Yalow was a talented, ambitious woman who overcame the barriers she faced at different stages of her life. Her challenges made her stronger. Her dedication to becoming a successful physicist was rare and unconventional in her time, and she wanted even more. She became a wife and a mother—in the traditional meaning of these words. She paid a price in human relations for her general uncompromising demeanor, but she succeeded spectacularly in her scientific achievements. This is why she felt justified to encourage her

[12] Roger Guillemin and Andrew V. Schally shared the other half of the 1977 prize for their discoveries concerning hormone production in the brain.

student audience about the opportunities for women in science, telling them, "You can have it all!"

Mildred Dresselhaus

Mildred Dresselhaus, 2002. Photograph by M. Hargittai.

"The Queen of Carbon Science," wrote *US News and World Report* in 2012,[13] after the announcement of the Kavli Prize for nanoscience, which was given to Mildred Dresselhaus (1930-2017). She received it "for her pioneering contributions to the study of... nanostructures." The Norwegian Academy of Science and Letters awards the Kavli Prize every other year for outstanding achievements in three fields: astrophysics, nanoscience, and neuroscience.

[13] M. Cimonds, "Queen of Carbon Science: Kavli Prize Winner is a Nanoscience Pioneer," *U.S. News & World Report*, July 27, 2012.

The Kavli Prize is still not well known—2012 was only the third time it was awarded—but its monetary weight is commensurate with that of the Nobel Prize. When I visited Dresselhaus in her office at MIT, I learned that in spite of her busy daily schedule, she found time to play chamber music with her friends almost every evening. When she was a child, she received a music scholarship and was thinking of a career as a musician, but then Paul de Kruif's book *Microbe Hunters* pushed her toward science. The final impulse came during her college years from a physics scholarship abroad that she could not miss, and she became a physicist.

Her parents were Jewish immigrants from Europe in the 1920s, her father from Poland and her mother from Galicia. She was born Mildred Spiewak in Brooklyn and started her university education at Hunter College, as many other gifted children of immigrant families did. At the time, the tuition-free Hunter College was the women's branch of the City University of New York.

After graduating, she received a Fulbright scholarship for a stay at the Cavendish Laboratory in Cambridge. Her PhD was from the University of Chicago. The great Enrico Fermi was still there, and she attended his course on quantum mechanics. Alas, not all the professors were like Fermi. Her faculty advisor thought that women should not go to graduate school; he considered it a waste of resources. Still, she persevered and did excellent research on superconductors, which was a hot topic in physics at the time.

She met her future husband, Eugene (Gene) Dresselhaus, while at the University of Chicago, and they married in 1958. He was a theoretical physicist and got a job at Cornell University. Cornell did not employ Mildred on account of the anti-nepotism rule, as was the case for other women included in this volume. Still, she stayed, supported by a two-year grant from the National Institutes of Health. After those two years the couple moved to the Lincoln Laboratory, a defense venue at MIT, where both could be employed. She became involved with magneto-optics, the study of the effect of electromagnetic waves on materials exposed to a magnetic field. She investigated graphite, and this choice determined her interests for a long time.

They had four children, born between 1959 and 1964. Her supervisor complained that she came in late each morning, so she took a year off. Luckily, after seven years at Lincoln Lab, she got a visiting professorship, still at MIT, which eventually led to a regular appointment. When, years later, she became the director of the Materials Center at MIT, her husband joined her laboratory. They published numerous papers together and co-authored books widely used by researchers in their field.

The Dresselhaus couple was rather unique in that Mildred was more successful and more famous than Gene, but this did not cause any conflict within the family. Managing the household with four children, performing high-level, frontier research, and serving in demanding positions of administration was possible only because she could rely on his support. He shared all her domestic duties and chores. Also, they had a babysitter, the same woman for 29 years.

Her most important research projects dealt with carbon, and specifically intercalation compounds. These are materials in which atoms of different elements or molecules are inserted between the layers of a lattice, which usually is graphite. She and her group were instrumental in the development of nanoscience, that is, the study of molecules and structures at the nanometer scale, between 1 and 100 nm.[14] She joined this field in the early 1970s. Soon after, her group became interested in fullerenes and nanotubes. In 1985, there was big news that a research group in Texas had found a new form of carbon, known as buckminsterfullerene, in short "buckyball," a beautiful symmetrical molecule consisting of 60 carbon atoms that looked like a soccer ball. By the time of its discovery the Dresselhaus group had been involved in related research for many years. She did not mention it, but it must have been on her mind that they could have been the ones to discover the "buckyball."

Two Nobel Prizes showed the importance of carbon research, but they both went to others, each for single milestone discoveries. One was in 1996 to Robert Curl, Harry Kroto, and Richard Smalley for the discovery of buckminsterfullerene, and the other in 2010 to Andre Geim and Konstantin Novoselov for their discoveries concerning graphene. Graphene is a two-dimensional carbon layer that can be looked at as a single graphite sheet or as an opened-up nanotube, which the Dresselhaus group had studied for years.

In 2012, Mildred Dresselhaus's contributions to carbon science were recognized by the Kavli Prize. This was one in a series of many prestigious recognitions she received over the years. They included the National Medal of Science in 1990, the Heinz Award in 2005, the L'Oréal UNESCO Award for Women in Science in 2009, and the Enrico Fermi Award in 2012, as well as memberships in the National Academy of Sciences (1985), the National Academy of Engineering (1974), and the American Academy of Arts and Sciences (1974).

[14] One nanometer (nm) is 10^{-9} meter.

After many years of being directly involved in scientific research, Dresselhaus took high-level administrative positions. She served as president of the American Physical Society, president of the American Association for the Advancement of Science, and treasurer of the National Academy of Sciences. For a while, she held the government position of director of the Office of Science at the Department of Energy. There, she was responsible for reversing the trend of decreasing support for the physical sciences in the federal budget. Serving in administration never entirely kept her from doing science.

As a scientist, she was also wholly dedicated to women's issues. This started with mentoring female students at MIT. In the 1960s, women were a mere 4 percent of the MIT student body. She was asked to help in the evaluation of undergraduate applications. In doing so, she realized that it was much harder for women to enroll at MIT than for men. Dormitory space was limited, and women did not usually perform as well academically as men due to social factors of discrimination and harassment. In the late 1960s, she prepared a motion to adopt equal requirements in admission, which was accepted. Later, she and other female colleagues started a women's forum to voice issues. By about 2000, the number of women at MIT had increased almost tenfold compared to what it was in the 1960s. In the late 1990s, another MIT professor, a biologist named Nancy Hopkins, looked into the question of women scientists at MIT and prepared a report, which has become known throughout the United States.[15]

Dresselhaus passed away on February 21, 2017. There were many obituaries, among them one in the *New York Times*. MIT's president L. Rafael Reif called her "a giant, an exceptionally creative scientist and engineer, and a delightful human being."[16] Among her many firsts, she was the first woman to become a tenured professor at MIT and the first woman to win both the National Medal of Science and the National Medal of Engineering.

Donna T. Strickland

Donna T. Strickland (b. 1959) made her dream come true when she became a physicist-engineer and worked with lasers. With her mentor, she invented

[15] "A Study of the Status of Women Faculty in Science at MIT," *MIT Faculty Newsletter* 11, no. 4 (March 1999), http://web.mit.edu/fnl/women/women.html.

[16] https://news.mit.edu/2017/letter-mit-community-regarding-death-mildred-dresselhaus-0221.

Donna T. Strickland. Source: Wikipedia.

a new laser capable of unprecedented intensity and precision. This chirped pulse amplification (CPA) earned the two a Nobel Prize.

Strickland was born in Guelph, Ontario, the middle child of an electrical engineer father and a teacher mother. On the day of her birth her father read in the daily paper that the first female engineer had graduated from the University of Toronto, the only woman in the class of 450. Her family was middle class and there was never any doubt that the children would go on to receive higher education.

Strickland went to local schools and by the time she got to the Guelph Collegiate Vocational Institute in grade 9, she knew she liked math and science. She was puzzled when a (female) teacher told her that math and science were for boys. Strickland was afraid that enjoying math and science would make her appear nerdy, but then found this not to be the case. When it came to deciding what to study after high school, she hesitated between engineering and physics, but happily McMaster University in Hamilton, Ontario, had an engineering physics program. She graduated from McMaster in 1981 with a bachelor's degree in engineering as one of three women in engineering physics. For graduate studies she was looking for a school with an emphasis on lasers and found two in the United States: the

Institute of Optics at the University of Rochester and the Optical Sciences Center of the University of Arizona. Rochester accepted her, and the very first week there she met Gérard Mourou (b. 1944). He was at the Laboratory of Laser Energetics, soon to be appointed professor; Strickland became one of his first PhD students.

Mourou gave Strickland a theoretical paper to read, which challenged her to build a laser that did not yet exist, a task worthy of a PhD project. This laser operated in the extreme ultraviolet region of the electromagnetic spectrum, meaning the irradiation would have very high frequencies at which existing lasers could not work. She needed a more intense laser than they had in Mourou's laboratory. When nothing else worked, Mourou suggested a new approach in which the frequency changes through the laser pulse, analogous to sound frequency changes in a bird's chirp. Hence they called it "chirped pulse amplification." The pulse would be stretched, chirped, and then amplified. This was the principle, but first it had to be built, and it took about a year for Strickland to do so.

The laser worked, and they thought others might come to the same idea, so they set about quickly getting the research published. They chose the periodical *Optics Communications* for its quick turnaround. Their paper was three pages long, fitting for this journal, which did not publish lengthy discussions. Whether it was their haste or an error in production, the article first appeared on October 15, 1985, featuring an incorrect figure. It was so important that rather than publishing an erratum, the entire paper was republished on December 1 later that year. The two authors appeared as Donna Strickland and Gerard Mourou and the title of the piece was "Compression of Amplified Chirped Optical Pulses."[17] It was Strickland's first publication and it was the basis of their Nobel Prize in 2018. Looking back, its impact was slow to take effect as certain necessary materials have only recently become available, such that even small academic laboratories are now able to build their own CPA systems.

This technical invention did not suffice for an entire PhD project; studies had to be conducted, measurements made, data analyzed, and everything had to be properly described. The writing was not yet completed even by the time Strickland had begun her postdoctoral fellowship at the physics division of the Canadian National Research Council in Ottawa.

[17] Donna Strickland and Gerard Mourou, "Compression of Amplified Chirped Optical Pulses," *Optics Communications* 55, no. 3 (1985): 219–221.

Strickland and her future husband met while still back in the Rochester laboratory. Doug Dykaar was a graduate student in electrical engineering. They were married in 1991 after she had completed her postdoctoral stint in Ottawa. He found a dream job at Bell Labs in New Jersey. She could not find anything of her own close by, only a job at the Lawrence Livermore National Laboratory in California, where she worked on developing new CPA lasers. After one year, she moved back east to be closer to her husband and worked at Princeton University as a member of the technical staff of the Photonics and Electro-Optic Materials Center. For her, none of these was a dream job, but they were very convenient for going through her pregnancies with Adam and Hanna. (The latter would go on to become an astrophysicist.) Finally, there was an opening for her in academia, and she became a professor at the University of Waterloo in Ontario, very close to where she grew up. This time, Doug followed her and found a job in industry. She has been happy teaching and doing research at Waterloo ever since. She has also been active in the Optical Society of America and the Canadian Association of Physicists, taking up leadership positions in both organizations.

In 2018, 33 years after the invention of the CPA laser, Strickland and Mourou were jointly awarded half of the Nobel Prize for Physics "for groundbreaking inventions in the field of laser physics." The specific motivation for Mourou and Strickland's half was "for their method of generating high-intensity, ultra-short optical pulses."[18] "Ultra-short" here means in the order of an atto-second, 10^{-18} of a second. The ultrashort high-intensity pulses generated by CPA have found applications in the diverse areas of medicine, industry, and defense, and the areas of applications keep expanding. The Nobel Prize brought with it other substantial recognitions. In 2019, Strickland was named Companion of the Order of Canada. In 2020 she was elected a member of the National Academy of Sciences of the U.S.A. and a Fellow of the Royal Society (London).

Donna Strickland concluded her autobiography on the Nobel site with this pledge: "I will use the platform the Nobel Prize affords me to continue to advocate on behalf of science and the many scientists who devote their lives to probing the fundamental questions."[19]

[18] The other half went to Arthur Ashkin (1922–2020) "for the optical tweezers and their application to biological systems."

[19] "Donna Strickland—Biographical," Nobel Prize, April 9, 2021, https://www.nobelprize.org/prizes/physics/2018/strickland/biographical/.

4
Crystallographers

X-ray crystallography is one of the great success stories in 20th-century science. It started with the discovery of X-rays by Wilhelm Conrad Roentgen in 1895. Soon, X-rays found eminent applications in medical diagnosis based on the differences in their penetration in softer and harder tissues. As they did not penetrate the bone, they could be used for diagnosing fractures. Subsequently, they proved invaluable in diagnosing tuberculosis as they were capable of detecting structural alterations in the lung.

In 1912, an entirely different use of X-rays came to the forefront. P. P. Ewald, a fresh theoretical physicist PhD at the University of Munich, in a discussion with one of the younger professors at the Institute of Theoretical Physics, Max Laue, raised the possibility of an interference experiment. They considered two suppositions: one was that X-rays have a wave nature—as they were by nature electromagnetic waves—and the other was that in crystals the atoms were in regular and periodic arrangement. These two suppositions are common knowledge today but had not been experimentally proven before 1912. Laue asked two young experimental physicists to shine a beam of X-rays onto a crystal sample. The resulting interference pattern was their proof. Seizing on this observation in 1913, the father-and-son team of W. Henry and W. Lawrence Bragg came up with the idea of using such experiments for the determination of crystal structures in order to determine the distances between the atoms—the building elements of crystals. This was the beginning of X-ray crystallography. Max von Laue[1] (1914) and the two Braggs (1915) received the Nobel Prize in Physics for their respective discoveries.

Over the years, women scientists contributed conspicuously to the spectacular success of X-ray crystallography. There were a number of factors at play. From early on, the principal male actors in the field simply did not object to women's active involvement. The relatively short time one had to spend in the laboratory doing the experiment and the longer periods

[1] Max Laue became Max von Laue when his family received nobility in 1913.

involving calculations, and later computations, was easier to fit into women's other obligations. Once there were some successful women in the field, they served as attractors or role models for budding female researchers. Here we introduce the reader to a few world-renowned crystallographers. We also present some pioneer contributors to the creation of data banks and show examples of the visualization of crystallography.

Kathleen Lonsdale

Detail of an exhibition tableau honoring Kathleen Lonsdale at the Royal Institution, London. Photograph by M. Hargittai.

Memorial plaque in Seven Kings, the Borough of Redbridge, London. Photograph by and courtesy of Steve Roffey.

Kathleen Lonsdale (née Yardley, 1903–1971), a pioneer crystallographer, was born in Ireland. She was a good student, especially in the sciences. She had to attend the science classes at the boys' high school, as such classes were not offered at the girls' school. With a scholarship, she studied at Bedford College for Women, part of University College London (UCL). Mathematics and physics were her principal subjects. As always, she excelled there as well and finished at the top of her class. The Nobel laureate physicist W. Henry Bragg invited her to his research group, which was busy determining the structure of small organic molecules.

She met her future husband at UCL, an engineering student. After their marriage, they moved to Leeds, where she continued working on X-ray diffraction in Christopher K. Ingold's group. Her most important scientific result dates back to this time, the determination of the structure of hexamethylbenzene as a *planar* hexagonal ring. Bragg supported her even though his earlier suggestion was a *puckered* ring for the molecule.

The Lonsdales moved to London in 1930, where her husband got a permanent job. They had three children, and for a few years Kathleen was occupied with bringing them up. In 1934, Bragg received funds and suggested she use some to hire childcare so that she could go back to work. This was a most forward-pointing approach. By then Bragg was at the Royal Institution, where Lonsdale stayed for 15 years. After the war she returned to UCL as a reader of crystallography, and in 1949 she was appointed professor of chemistry and head of the Department of Crystallography. She was UCL's first tenured female professor.

She received many honors and broke several records for women scientists. In 1945, she was one of the first two women to be elected a Fellow of the Royal Society (the other was the biochemist Marjory Stephenson). In 1956, she became Dame Commander of the Order of the British Empire. In 1966, she was elected as the first female president of the International Union of Crystallography. *Lonsdaleite*, a rare form of diamond found in meteorites, immortalizes her name.

Dorothy Crowfoot Hodgkin

Dorothy Mary Hodgkin (née Crowfoot, 1910–1994) was born in Cairo into a well-established British family serving the British Empire. From an early age, her father's postings made it possible for her to gain an international outlook and to get acquainted with ancient civilizations, as well as with

Dorothy Hodgkin on a British postage stamp, 1996.

Dorothy Hodgkin on a British postage stamp, 2010.

grave social problems in a variety of locations. She had two younger sisters whose care at times fell on her when in England, while their parents were away for lengthy periods. Her circumstances fostered the development of her independence and decision-making ability.

She had a thirst for knowledge. During her high school years, she performed homemade experiments using ingredients purchased from local stores. She noticed that boys received more serious education in science and mathematics than girls, and so she joined the boys' classes. She received encouragement from the fact that the chemistry teacher was a woman and from reading the lectures on science for children developed by the two Braggs.

In 1928, Dorothy Crowfoot began studies at Oxford University, majoring in chemistry. Oxford had been allowing women to study there for about 50 years by that time, but it had been less than a decade since it began granting them degrees. Women were still vastly underrepresented among the student body, and many people still held the view that their presence had a demoralizing influence. Various societies, among them a prestigious student society for chemists, would not accept female members even then. For the best women students and their tutors, all of these difficulties only enhanced their determination to excel. Dorothy went on to receive excellent instruction from world authorities in chemistry, several of them Nobel laureates.

Following graduation, she moved to Cambridge and worked with the crystallographer J. Desmond Bernal for two years, 1932–1934. He was a visionary, one of the greatest scientists of the 20th century. She came away from her Cambridge period with an impressive list of discoveries and publications. One of her papers reported on the first X-ray diffraction experiment on a protein. This was the first indication that the structures of these vitally important substances could be determined. A crucial difference was observed in their experiments between the diffraction patterns of dry and wet protein crystals. The presence of the liquid from which the crystal was grown—called the "mother liquor"—was essential to produce a usable diffraction pattern. Thus, the conditions were set for protein crystallography. Bernal and Hodgkin remained lifelong friends and trusted colleagues, but they each followed their own scientific path. Hodgkin spent the rest of her professional career at Oxford. Bernal too left Cambridge and was appointed to a professorship of crystallography, the first such appointment ever, at Birkbeck College, UCL.

Oxford and Cambridge provided the best of professional environments, but not the friendliest to aspiring women. Although women could be students and researchers at Oxford, with the same rights as their male counterparts, socially this was not quite the case. For example, the prestigious Alembic Club did not accept women as members until 1950. As late as 1949, when she marched into an Alembic meeting, Dorothy was forced out, physically. She was most successful, though, in managing the tensions between the path of a dedicated researcher and the life of a devoted wife and mother.

In the mid-1930s, she met Thomas Lionel Hodgkin (1910–1982), an Oxford graduate from a distinguished family of scholars, professors, and physicians. He was a Marxist who became a respected scholar at Oxford specializing in African politics and history. They married in December 1937, and following a brief honeymoon, she returned to her laboratory. She was the first Oxford Fellow to give birth while in position. Paid maternity leave was still decades away, but Somerville College proved to be enlightened and gave her paid leave. She was not quite ready to keep house, let alone tend to children, and Thomas was away most of the time. He did not make much of an effort to carry any of the burden at home, but friends and neighbors as well as her mother proved most helpful. In time, the Hodgkins could afford to employ help for the children and someone for cooking and housekeeping. In 1955, she fought the university to provide equal pay for male and female graduate students. When a female student got married, the practice used to be to cut her grant.

Dorothy Hodgkin barely missed anything in her work due to her family life. She excelled in the determination of biologically important structures. Most conspicuous among them was penicillin, as it had great strategic war-time importance. By 1939, she had created such a name for herself in crystallography that nobody was surprised when she undertook this task. Even compared with the number of soldiers killed in battle, the number of those lost to bacterial infections was significant. In order to attempt the synthetic production of penicillin, its composition and structure had to be known. Hodgkin and her associates worked it out. When Bernal told her in 1945 that she would one day receive the Nobel Prize for this, she told him that election to Fellow of the Royal Society (FRS) would be more important to her. As it happened, she was elected FRS in 1947, whereas the Nobel Prize came her way many years later, in 1964. By the time she was elected FRS, she had already given birth to three children. In the short term, Hodgkin's discovery of the penicillin structure did not contribute to the mass production of penicillin; this was solved by fermentation technology. In the long run, however, her results amply contributed to the success of the pharmaceutical industry in producing a good number of antibiotics.

Hodgkin produced further world-class discoveries in X-ray crystallography. Suffice it to mention the determination of such biologically important substances as Vitamin B_{12} and insulin. Both took many years to finally yield the right structure. At the time, these achievements were at the frontier of science. Alas, Oxford did not seem to appreciate her or the value of her work. When the possibility of establishing a position of reader in Crystallography

Part of a large group of participants at the 1934 International Conference on Physics in London: Dorothy Crowfoot (as Hodgkin was then) is second from the left, back row; Patrick Blackett is in front of her to the right, and J. Desmond Bernal is in front of Blackett to the right. Photograph by London Panoramic Company. Source: Wikimedia Commons.

opened up, it went to someone else. But even her appointment to the relatively low position of demonstrator was a step forward. Her disadvantaged situation was reflected not only in her remuneration but in the limited possibilities she had to employ students and assistants and expand her inadequate lab space. In contrast, for decades she enjoyed generous support from the Rockefeller Foundation. The Royal Society also acknowledged her achievements; in 1956 she was awarded its prestigious Royal Medal, and in 1960 the Society decided to award her with a recently established Royal Society Professorship endowed by Lord Wolfson. For accepting this honor, however, she had to have the consent of Oxford University. One would have thought it should have been granted enthusiastically as it meant both prestige and savings (her salary) for the university. But it was only after some negotiations that Hodgkin could finally accept; Oxford would even grant her the title of professor. In 1976 the Royal Society awarded her its highest recognition, the Copley Medal, first given in 1731, and she has remained the only woman with this distinction.

Hodgkin received the Nobel Prize in Chemistry, unshared, in 1964. The citation credited "her determination by X-ray techniques of the structures of important biochemical substances." If there was ever a Nobel laureate whose demeanor the Nobel Prize did not change, it was Hodgkin. She continued to be the modest, unassuming Dorothy, interested in the affairs of others and ever helpful. No wonder she was so popular among her friends and peers and the members of the international community of crystallographers. Even when she no longer carried out research, she attended the international congresses of crystallography until she no longer could. The one in Beijing in 1993 was her last. She was wheelchair-bound but still could bask in the love and care of her colleagues and in the attention of younger crystallographers who had known her only from the literature. She died the following year.

Her tributes did not stop with her passing. For example, she appeared on a British stamp soon after her death, in 1996, along with a few other women celebrities. Then, in 2010, another Dorothy Hodgkin stamp was released as part of the 350th anniversary celebrations of the founding of the Royal Society. She was the only woman depicted among the 10 illustrious Fellows. Listing the other 9 adds to the weight of this distinction: Isaac Newton, Edward Jenner, Joseph Lister, Benjamin Franklin, Charles Babbage, Robert Boyle, Ernest Rutherford, Nicholas Shackleton, and Alfred Russel Wallace.

Dorothy Hodgkin has been the subject of artistic expression. Patricia Fara has written about two of the pictures of Hodgkin.[2] Both emphasize her severely deformed hands, the consequence of rheumatic arthritis she suffered since her youth. A drawing of Hodgkin's hands by Henry Moore (1978) is at the headquarters of the Royal Society of London. An oil painting of Hodgkin by Maggi Hambling (1985) is displayed in the National Portrait Gallery in London. This image shows Hodgkin at her desk with two pairs of hands, as if in redoubled activity despite her handicap.

The Royal Society has another oil painting of Hodgkin, this one by Bryan Organ (1982). The Council of the Royal Society decided in 1977 that a woman Fellow should finally have a portrait, though it offered no sponsorship for this.[3] A brief and most successful public subscription brought together the

[2] Patricia Fara, "Pictures of Dorothy Hodgkin," *Endeavour* 27, no. 2 (2002): 85–86.

[3] John Cornforth, "A Memory of Dorothy Hodgkin," *The Chemical Intelligencer* 4, no. 4 (1998): 57–58 and "Portrait of Dorothy Hodgkin," *Notes and Records of the Royal Society of London* 37, no. 1 (1982).

necessary Portrait Fund. Next, they had to find an artist, Graham Sutherland OM (1903-1980). As it turned out, Hodgkin and Sutherland had met in person at what must be one of the world's most exclusive luncheons. It is given once every five years by the Queen for the members of the Order of Merit, OM, whose membership is limited to 24; both Hodgkin and Sutherland were members. Sutherland made two sketches, one of which is at the Royal Society and the other at the Science History Institute in Philadelphia. These sketches eminently grasped Hodgkin's demeanor. Alas, Sutherland fell ill and died before he could complete the painting. The task was then taken up by Bryan Organ.[4]

Rosalind Franklin

Rosalind Franklin (1920-1958) was born into an affluent Anglo-Jewish family. She had a carefree childhood, received a good education, and showed interest in science. She was good in languages and loved sports. Her devotion to mountain-climbing originated from repeated visits to Norway. She enrolled at the University of Cambridge in 1938 and was a

Rosalind Franklin's portrait in a window display at King's College London. Photograph by M. Hargittai.

[4] Michal Meyer, "A Drawing of a Biochemist Connects Two British Political Icons," *Distillations*, March 17, 2018.

Plaque on the house where Franklin lived, at Donovan Court, Drayton Gardens, in London. Photograph by M. Hargittai.

student of Newnham College. She earned her bachelor's degree in physical chemistry in 1941.

A research scholarship helped her to stay at Cambridge and carry out research on polymerization with the future Nobel laureate Professor R. G. W. Norrish. Unfortunately, the project and their interactions gained little traction, and she gave up the rest of the scholarship. She moved on to the British Coal Utilization Research Association, where her work helped the war effort. Her project was interesting and she was able to complete her PhD dissertation simultaneously, receiving her doctorate in physical chemistry in 1947. Between 1947 and 1950, she worked in a French state chemistry laboratory in Paris, the Laboratoire Central des Services Chimiques de l'État. This period was pleasant and fruitful. She developed her skills in X-ray crystallography for structure determination and combined it with model building. She investigated the structure of partially crystalline carbon, a research direction at the frontier of the then accepted possibilities of X-ray crystallography.

Next, she embarked on the structure determination of DNA that was even more challenging in testing those limits. This was at King's College, University of London, where they had a biophysics unit. It was organized and headed by the physicist John Randall, who met Franklin through the renowned Oxford mathematical chemist Charles Coulson. Another scientist in Randall's laboratory, Maurice Wilkins, had already been engaged in the X-ray diffraction studies of DNA, and Randall had failed to clarify the two positions of Wilkins and Franklin. Franklin had become an independent researcher, whereas Wilkins was under the impression that she was hired to work for him. Add to this the disadvantaged setup for female associates at King's College in general, and the situation was ripe with potential conflicts.

Franklin worked with a graduate student, Raymond Gosling (1926–2015). The two together did excellent experiments and produced excellent photographs of the X-ray diffraction patterns of DNA. The best of them was photograph No. 51, which would become famous. Wilkins continued his meticulous X-ray diffraction studies of DNA and interacted with the other British group working on the DNA structure, Francis Crick and James D. Watson at Cambridge's Cavendish Laboratory. Crick was working on his PhD and Watson was an American PhD. At King's College, Franklin and Wilkins were unable to create a tolerable relationship, and the situation was fast deteriorating. At one point the angry Wilkins showed the visiting Watson Franklin and Gosling's photograph No. 51 without Franklin's knowledge. This photograph was a crucial piece of evidence for the double-helix structure of DNA. As a consequence of Wilkins's clandestine act, Watson and Crick could neither mention nor thank Franklin in their seminal publication. Franklin never learned about Wilkins's action behind her back.

In 1953, Watson and Crick, Franklin and Gosling, and Wilkins and his two associates published their respective results about the DNA structure in three short communications, appearing back to back, in *Nature*. Aaron Klug, who worked with Franklin on virus structures and subsequently became a Nobel laureate in 1982 for his discoveries in electron crystallography, years later examined Franklin's laboratory journal. He concluded that she was much further ahead in her work than had been supposed. Had Watson and Crick not made the discovery of the double-helix structure, she might well have arrived there independently. It would have come out in steps rather than in one stroke, as it did in Watson and Crick's report. Accordingly, the impact would have been more gradual, but the consequences would have been the same.

Franklin decided to leave King's College and her impossible situation there, and move to Birkbeck College at J. Desmond Bernal's invitation. There, her principal research project was the structure of the tobacco mosaic virus, consisting of a protein molecule enveloping a nucleic acid, the carrier of genetic information. This simple virus provided an excellent case for studying the interactions of proteins and nucleic acids. She applied her previous experiences in coal research and the DNA study, and produced results of lasting value. The creative and friendly environment of Birkbeck was conducive to a most productive final period of Franklin's life before she succumbed to a devastating illness. Klug joined her in this work, and many of their findings about virus structures have remained valid to this date.[5]

Plaque at King's College, London, commemorating the X-ray diffraction studies of DNA by R. E. Franklin, R. G. Gosling, A. R. Strokes, M. H. F. Wilkins, and H. R. Wilson. Photograph by M. Hargittai.

Crick, Watson, and Wilkins were awarded the Nobel Prize in Physiology or Medicine in 1962 for the double-helix structure of DNA. It has been much debated whether Rosalind Franklin should have been among the awardees. However, this is not a relevant issue. There is no posthumous Nobel Prize and the awardees did not receive nominations while Franklin was still alive. It is a negative feature of the Nobel that it often appears to be a watershed as it immortalizes the laureates and helps us forget those omitted. The Swedish stamp commemorating the 1962 Nobel Prize in Physiology or Medicine is emblematic of this unfairness. The names of the three laureates appear against the backdrop of two characteristic images of the discovery. One is the famous Franklin-Gosling X-ray diffraction pattern of photograph No. 51, and the other is the Watson-Crick model. The

[5] Aaron Klug mentioned in his Nobel lecture on December 8, 1982, that Rosalind Franklin set him "the example of tackling large and difficult problems. Had her life not been cut tragically short, she might well have stood in this place on an earlier occasion." *Nobel Lectures, Chemistry, 1981–1990* (Singapore: World Scientific, 1992), 79.

names of Franklin and Gosling are missing, so a casual observer would easily assume that the diffraction pattern was also the result of the work of the laureates.

That Franklin's (and Gosling's) achievements did not disappear into oblivion is largely due, however paradoxically, to Watson's negative portrayal of Franklin in his bestselling book, *The Double Helix*. This triggered a second look at Franklin's contributions to the discovery, which led to her universal recognition. By now, another paradox has developed in the recognition, or lack of thereof, in the history of the discovery of the double helix. As Franklin's recognition has risen, Wilkins's has plummeted. Nobel Prize or not, Franklin is probably more recognized than Wilkins. Wilkins worked at King's for decades, whereas Franklin's tenure there was considerably shorter, yet even at King's their level of recognition is the same. There is now a Franklin-Wilkins Building at King's College, housing the Franklin-Wilkins Library. The plaque commemorating their achievements on the façade of the college displays their names and the names of their associates with equal weight. Franklin's name and achievements have been commemorated in numerous other venues. Just one example is the Rosalind Franklin University of Medicine and Science in North Chicago, Illinois, renamed for Franklin in 2004. The school's logo displays a schematic image of the Franklin-Gosling X-ray diffraction pattern. In Franklin's case, her conspicuous lack of Nobel recognition did not lead to her disappearance from memory. On the contrary, it has enhanced the awareness of her plight as a woman scientist and of the importance of her discoveries. Her example continues to inspire and encourage young women to embark on a career in science, pointing to the hurdles that gifted and assertive young women may meet along their path in the sciences. It calls convincingly for the need to improve the working environment. Indeed, six decades since Franklin's passing, there is still a great deal to achieve in this respect.

Isabella Karle

Isabella Karle (née Lugowski, 1921–2017) and her husband, Jerome Karle (1918–2013), represent a shining example of a "scientific couple." The two Karles spent all of their scientific lives together, working on the same or related projects. In addition, each had their own research line and achievements. Isabella and Jerome met at the University of Michigan and did

Isabella Karle, around 1955. Courtesy of Isabella Karle.

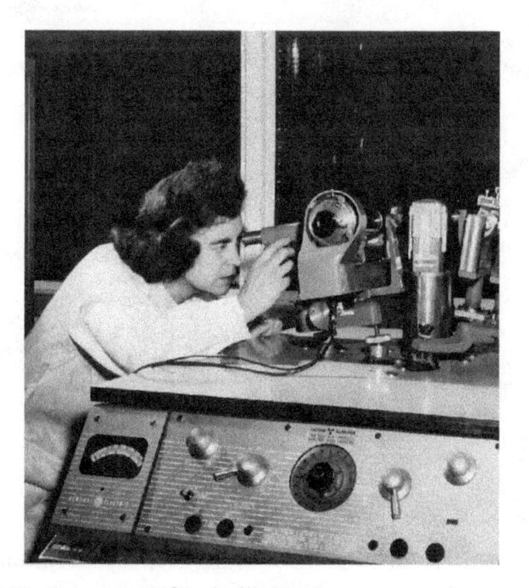

Isabella Karle, 1960. Courtesy of Isabella Karle.

their PhD studies with Lawrence Brockway in gas-phase electron diffraction. They married at the end of the second academic year.

Isabella was born in Detroit, Michigan. Her father was a housepainter and her mother a seamstress; both were immigrants from Poland. Isabella learned English when she started school. During her high school years she decided to study chemistry. After Isabella and Jerome graduated from the University of Michigan they both participated in the Manhattan Project at the University of Chicago. They were involved in finding procedures to produce fission fuel plutonium without impurities. After the war, they stayed at the University of Michigan for two years. There was such a shortage of instructors that the university was happy to have them both in spite of the anti-nepotism rule. Then, in 1946, they both got positions at the Naval Research Laboratory (NRL) in Washington, D.C., where they were officially allowed to work together.

According to Isabella, they "worked together separately," that is, they complemented each other. He was theory-oriented and she often provided experimental support for his ideas. In their doctoral studies, they improved the gas-phase electron diffraction technique for the determination of molecular structure.

The Karles (Jerome and Isabella in the middle) and the Hargittais (Magdolna and Istvan) at a scientific meeting, 1978, in Pécs, Hungary. Photographer unknown.

By improving and better understanding the experiments, they made it possible to determine finer details of the molecular structure. The structure of a molecule can be imagined by a geometrical model in which each atom of the molecule is represented by a small sphere; these spheres have fixed positions in three-dimensional space. This does not mean the atoms are rigid; in fact, they are in perpetual motion about these fixed positions. The improvements Jerome and Isabella introduced uncovered the synergy between this motion and the geometry of the molecule.

In the early 1950s, Jerome and the mathematician Herbert Hauptman together developed the so-called direct method for the analysis of X-ray diffraction data. They proposed a technique and made it possible to determine larger structures than before, and to do so more accurately. The crystallographers, whose work was to benefit from these new techniques, were slow to accept them. Eventually, it was Isabella who gave a real push for the acceptance of the techniques. She successfully investigated complex molecules consisting of 40 to 60 atoms, which up until that time could not reliably be done. She worked out the connection between the mathematical description of the new technique and her X-ray diffraction data. This, finally, convinced the crystallographers, and thus she had a major role in making the direct method a successful tool in X-ray crystallography. This discovery eventually brought the Nobel Prize to Herbert Hauptman and Jerome Karle.

In 1985, on the day of the announcement of the Nobel Prize, the Nobel Committee wanted to call Jerome to inform him of the great news. He was just flying back home from Europe. The Committee called the captain of the plane, who then went out into the cabin and told Jerome the news. Jerome told me later that his first thought was whether Isabella was included in the prize, but the captain did not know. The remaining hours of the flight were agonizing for Jerome. When he got home, he learned that she was not included, and the sadness he felt over this never left him. Many of their colleagues shared the belief that Isabella should have been included in the prize. She herself felt that it would have been wonderful to receive the prize together with her husband, but many other prestigious awards consoled her.

After proving how useful the direct method was, her attention turned increasingly toward the structure determination of large, biologically important molecules. She uncovered details about the structures of peptides, steroids, and alkaloids, and her results advanced chemical and biochemical research all over the world.

The Karles had three daughters, and Isabella considered herself lucky in that crystallography is a scientific field in which she could both do the science and bring up the children: "Crystallography wasn't something that you had to watch all the time. You could take it home with you, you could think about it while minding the babies. Most of the projects in crystallography, for example, start with an idea or a substance, and there is an end when you get the crystal structure. This is an isolated thing and in order to have a research project you want a number of related things that would go together, but it was possible to do it stepwise. In other projects there may be so much more interaction."[6] One of Isabella's most prestigious awards was the Aminoff Prize (1988) for pioneers in crystallography, likewise administered by the Royal Swedish Academy of Sciences. Some of her other distinctions: she was a member of the National Academy of Sciences (1978) and the first woman to receive the Bower Award and Prize for Achievement in Science from the Franklin Institute (1993), "for determining three-dimensional structure of molecules with X-ray diffraction." In 1995, she received the National Medal of Science from President Bill Clinton.

In 2009, after more than 60 years (a combined 127 years) of service, Jerome and Isabella Karle retired from the NRL. Secretary of the Navy Ray Mabus presented them with the Distinguished Civilian Service Award, the highest award a civilian can receive from the navy. Nothing illustrates better how long ago it was that they joined NRL than realizing that "[w]hen Jerome Karle began work at NRL, Franklin D. Roosevelt was president, gas was 21 cents a gallon, minimum wage was 30 cents per hour, and a first class postage stamp was 3 cents."[7]

A crucial question for the woman in scientific couples is whether she receives proper credit for her work. When I asked Isabella whether it ever happened that Jerome got the credit for something she did, she answered, "I suppose so." Did it happen the other way around? "Not often." Isabella and Jerome Karle remain a wonderful example of a scientific couple; their joint output was much more profound for having worked alongside one another rather than separately.

[6] Istvan Hargittai and Magdolna Hargittai, "Isabella L. Karle," in *Candid Science,* vol. 6: *More Conversations with Famous Scientists* (London: Imperial College Press, 2006), 408–409.

[7] McKinney, Donna, "Jerome and Isabella Karle Retire from NRL Following Six Decades of Scientific Exploration." Archived September 27, 2011 at the Wayback Machine, United States Naval Research Laboratory press release dated July 21, 2009.

Pioneers of Data Banks

Two women scientists figure among the pioneers of data banks. They became engaged in collecting, critically evaluating, storing, and making available valuable structural information. Barbara Mez-Starck did this for gas-phase molecules and Olga Kennard for crystal structures.

Barbara Mez-Starck

Barbara Mez-Starck in 1999. Courtesy of Natalja Vogt.

Barbara Starck, later Mez-Starck (1924–2001), had a carefree childhood in Germany. The marriage of her parents had a romantic beginning. Her father, Hermann C. Starck, was a German industrial magnate and her mother, the Hungarian Jewish Klára Sarkadi, was one of his many employees. Hermann C. Starck had rigorous rules that precluded any close connection between employer and employee. When one day he fired Klára Sarkadi, she was devastated, as she had always worked to the full satisfaction of her superiors. Devastated, for just a moment; in the next, Hermann proposed to her. After Barbara, there was one more child, her younger brother, Gerhard.

The idyllic life of the family changed when the Nazis came to power in 1933. The mother survived by hiding and bribes. The father spent a fortune to ensure a close to normal life for the children, and for a while he succeeded. When Barbara graduated from high school she started her studies in chemistry at Freiburg University, but she was excluded in 1942 when all "half-Jewish" students and workers were expelled from the universities. She found employment in a plant working with metals, and she did analyses in a chemical laboratory. All four members of the family survived, but all of Barbara's maternal relatives were murdered in Auschwitz.

After the war, Barbara continued her studies at Freiburg University. She earned her diploma (master's degree equivalent) in chemistry, and in 1959, her doctorate. She stayed at the university and worked as the assistant to a professor. He charged her with collecting the data he wanted to use in his lectures. She created a user-friendly database that other professors were also free to use, which they did. When Starck's professor died unexpectedly, the university decided that she should continue her activities, useful as they were. Her job was converted into one that officially prescribed her duties as a database developer. This included all data resulting from the application of all physical, theoretical, and computational techniques producing information on molecular structures in the gaseous phase. She was not an expert in all fields, but she created a network of international experts that covered everything belonging to the scope of her activities. The project was her life, and her marriage to Erwin Mez, when she was 50 years old, did not change her dedication to this professional path.

Her active participation in the project suffered a major blow in 1983, when she had a stroke which left her partially paralyzed and in great pain. Still, she carried on for years, though everything that she did required a huge effort. Fortunately, by then she had taken care to develop a successor to be in charge of the operations. For Starck, the next two decades were torture, but she could still oversee the stability and survival of her creation. To the end, she continued gathering and evaluating data. For the last 15 years she worked for free, but this did not bother her; she was independently wealthy. She spent as much money as needed for her treatment, whatever money could buy, which afforded little relief, despite the high cost. She initiated a foundation in order to provide the data bank a solid financial background. She was always a person of action, and no hardship could prevent her from reaching her goals. Indeed, hardships shaped much of her life and personality.

She was not interested in awards or distinctions, and never received any. She made no effort to immortalize her contributions. Nonetheless, her center for structure documentation can be considered her memorial, and it would be most appropriate to call the institution she created the Barbara Starck Center for Structure Documentation. She had no such ambitions. She preferred to withhold all traces of herself from her creation despite all the energy, time, and resources she gave to that center. She took care of the future of the data bank, but willed to have all her personal affairs there destroyed.

Olga Kennard

Olga Kennard in Cambridge, UK, 2003. Photograph by M. Hargittai.

Olga Kennard (née Weisz, b. 1924) was born in Budapest. Her father and uncle jointly ran a private bank. Her mother's family had long been involved in the business of antiques, and Olga learned a great deal about old cultures. The family was Jewish, and the growing anti-Semitism in Hungary and the ever harsher anti-Jewish legislation gave ample warning for them to organize their departure while it was still possible. They left for England; most of their family members who stayed in Hungary perished in the Holocaust.

When they left Hungary in August 1939, Olga was 15 years old. She hardly spoke English, but within weeks she had to start attending school. When she began at the Hove County School for Girls, her teachers wanted to test her English. They gave her a story to read, which turned out to be the English translation of a Latin fable. She recognized it from her Latin studies back in Budapest. Thus she demonstrated sufficient proficiency and was placed in the matriculation class. At the final exams, again she was lucky because she could answer questions about Shakespeare although she would not have been able to buy anything in a shop because she did not know the spoken language. Next, she went to a mixed school in Evesham, where she was the only girl in her class. She was ambitious and persuaded the headmaster to let her sit for the examination that would allow entrance to Cambridge—no one from that school had yet got in; she got in.

What pushed her toward science was its promise of permanency, where there would be no nulls or voids between her Hungarian background and what she was learning in English schools, as there would be with history and other humanities subjects. She started in the natural sciences at Newnham College, a women's college. At that time, women did not get a degree, only a certificate indicating eligibility for a degree had they been men. Olga and others received their degrees in a belated ceremony, about 50 years later. Meanwhile, she studied chemistry, physics, crystallography, and mathematics, and in due course she earned her PhD.

She joined Max Perutz at the Cavendish Laboratory, whose principal research project was the determination of the structure of hemoglobin. In 1948 she married and changed her name to Kennard. She moved to London with her medical scientist husband and joined the Vision Research Unit of the Medical Research Council. The head of the unit was an eccentric professor, Hamilton Hartridge, who gave her impossibly large systems to solve, but when she failed, he was most encouraging. In 1961, she returned to Cambridge and stayed with the Chemistry Department until her retirement. She determined crystal structures using X-ray crystallography, among them some of the most important molecules in the living organism. One of them was adenosine triphosphate, which is also called the energy currency of the organism.

The Kennards had two daughters, but the marriage did not last. Olga brought up the girls by herself. Both became professionals, married, and had children. Olga remarried in 1994, to Sir Arnold Burgen (1922–2022), professor of pharmacology. Her greatest feat was the establishment of the Cambridge Crystallographic Data Centre.[8] It is a nonprofit organization that contains the Cambridge Structural Database, collecting and critically evaluating all crystal structure determinations of organic substances (those containing carbon) from all over the world. The database contains the structural information for well over a million substances and rapidly grows to this day. Once the data are accepted, they are made available for all researchers who live in a subscribing country, as most do; the fees are established in such a way that even poor nations can afford to participate.

Today, the Cambridge Crystallographic Data Centre is a world-renowned institution and its headquarters are in an award-winning building designed specifically for it. It all started humbly in the 1960s at the initiative of the pioneer crystallographer J. Desmond Bernal. Following Bernal's dream and supported by the British government, Kennard created something unique and lasting. Its main value is in the enhanced reliability of structural information originating from so many laboratories in the world. Their data have to correspond to certain quality requirements before being accepted and available to others.

Olga Kennard was elected Fellow of the Royal Society in 1987 and was made Officer of the British Empire the following year. In 2020, she received the Ewald Prize from the International Union of Crystallography for her outstanding contributions to crystallography. This is the highest recognition by the community of crystallographers and one of the highest in the field of crystallography.

Ada E. Yonath

Ada E. Yonath (née Livshitz, b. 1939) is an Israeli biochemist who has always loved challenging tasks. For quite some time she had been expected to win the Nobel Prize, and eventually it finally happened. "Nobel Prize for the chemistry of life," announced the BBC in October 2009. It was awarded jointly to Venkatraman Ramakrishnan, Thomas A. Steitz, and Yonath for the elucidation of the structure and function of the ribosome. It was a logical continuation of previous major discoveries. It started with Darwin's general theory of evolution, followed by the discovery of the double helix structure

[8] Cambridge Crystallographic Data Centre

Ada Yonath in Budapest, 2002. Photograph by M. Hargittai.

of DNA (1962), then the understanding of the copying mechanism of nucleic acids (2006), and finally, by Yonath and the others, with the demonstration of how the genetic code manifests not only as muscles, bones, and skin but also as thoughts and speech.

The ribosome is a giant system that can be described as the cell's "protein factory," which synthesizes proteins. It translates the genetic information carried by the DNA with the help of another nucleic acid, the messenger-RNA, for the production of proteins. The proteins are produced with fantastic speed in the ribosome. In them, the amino acids are linked to each other by peptide bonds. To produce such a bond in the laboratory may require extreme experimental conditions and considerable time. In contrast, the ribosome does this in microseconds and under mild conditions within the living cell. All cells of all living organisms, starting from the simplest bacteria, contain ribosomes, and, accordingly, they represent an obvious target for drugs. Thus, understanding their structure and function should help in drug design.

Yonath is a strong-willed, dedicated woman. Without these traits she could not have reached her goals. She was born in Jerusalem. Her father was a rabbi, and her parents immigrated to Palestine from Poland just after the Nazis came to power in 1933. Ada was 11 years old when her father died, and she had to help her mother support the family. She tutored younger children and did other chores. During her university studies she became interested in biochemistry and biophysics, receiving her master's degree in the latter. She worked toward her doctorate at the Weizmann Institute in Rehovot, on the structure of collagen. After she received her PhD, she continued with post-doctoral studies in the United States. First, she was at the Mellon Institute in Pittsburgh, then she joined the group of F. Albert Cotton at MIT, doing protein crystallography.

After two years she returned to Israel and started to build up her own group in protein crystallography at the Weizmann Institute. It took half a decade before things started working. Beginning in the early 1970s, protein crystallography in Israel was gradually moving to the international front lines. She collaborated with a number of laboratories, among them with H. G. Wittman of the Max Planck Institute for Molecular Genetics in Berlin.

The major difficulty was to grow crystals that can be put into the X-ray machine. She saw that the Planck Institute had active and pure ribosomes in relatively large amounts from bacteria, so she suggested using it for crystallization. Other prominent scientists had failed to crystallize ribosomes, and this only enhanced the challenge she felt, although her attempts were followed with much scrutiny and skepticism. For years, Yonath and her colleagues tried all sorts of methods to stabilize the ribosome and produce good-quality X-ray diffraction patterns from them. They examined the ribosomes of bacteria that live under hard conditions, supposing that their ribosomes might be more sturdy to survive all the manipulations they had to apply. Having this in mind, they took bacteria from hot springs and from the Dead Sea, full of salt. By the early 1990s, they managed to prepare high-quality samples. They recorded nice diffraction pictures, but due to the size of the ribosome, the interpretation was extremely difficult.

More groups entered this area of research, and a race developed. Thomas Steitz's group at Yale University published the first X-ray structure of the ribosome in 1998, but without the atomic positions because of the low resolution of their data. Finally, three groups published the most important results, and the three leaders of these groups shared the 2009 Nobel Prize. It was awarded jointly to Venkatraman Ramakrishnan of Cambridge University, Thomas Steitz, and Ada Yonath "for studies of the structure and function of ribosome."

Gradually, as a result of the three groups, it became clear how the ribosome operates. The proteins are produced in the large subunit; this happens extremely quickly. The small subunit translates the information, transmitted by the messenger-RNA from the DNA, into the "language of proteins."

Alongside her work at the Max Planck Society, Yonath spent her entire career at the Weizmann Institute. In the late 1980s she became the director of the Mazer Center for Structural Biology and of the Kimmelman Center for Biomolecular Assemblies. She was recognized with memberships in science academies, honorary doctorates, and prestigious awards even prior to the Nobel Prize, including the Israel Prize (2002), the Wolf Prize (2006), the Rothschild Prize (2006), and the L'Oréal-UNESCO Award for European Woman in Life Science (2008). Years after the Nobel recognition, she was elected Foreign Member of the Royal Society (London, 2020).

She had a difficult career but did not experience any specific disadvantage because she was a woman scientist. After Yonath divorced, she sometimes felt that she was not always there when her daughter needed her. Her daughter had to be self-reliant and became independent and responsible from a very early age.

Caroline H. MacGillavry and the Art of M. C. Escher

Caroline H. MacGillavry (1904–1993) lived her whole life in Amsterdam. She studied chemistry, and following the example of her mentor, one of the pioneers in modern crystallography, Johannes M. Bijvoet (1892–1980), she became a crystallographer. She excelled in the determination of ever more complex molecular systems. For her achievements, in 1950, she was elected to the Royal Netherlands Academy of Arts and Sciences, the first woman accorded such an honor. She was also appointed to a professorship at the University of Amsterdam. She was married to J. H. Nieuwenhuijsen, an otolaryngologist.

MacGillavry recognized early on the relevance to crystallography of the art of the Dutch graphic artist M. C. Escher (1898–1972). She published a book, *Symmetry Aspects of M. C. Escher's Periodic Drawings*,[9] which became a pedagogical tool in crystallography. It attracted international attention to Escher's periodic drawings at a time when the artist was not so widely recognized as he would come to be.

[9] Caroline H. MacGillavry, *Symmetry Aspects of M. C. Escher's Periodic Drawings* (Utrecht: Bohn, Scheltema and Holkema, 1976).

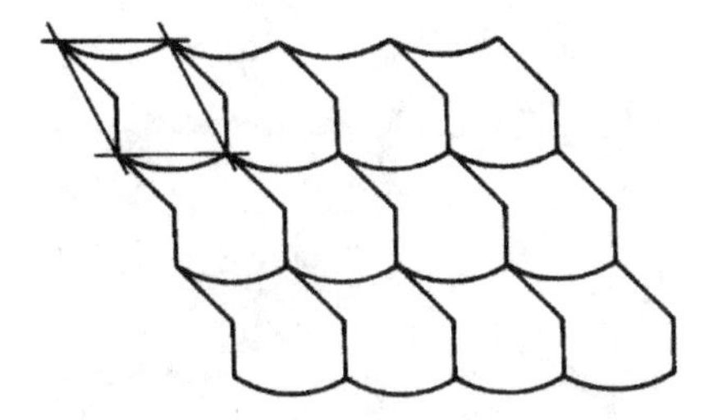

A simple two-dimensional pattern of a repeating motif. The box superimposed on the upper left of the pattern shows how the repeating motif was created on the basis of the original rectangular network.

A simple two-dimensional pattern of the repeating motif of a fish and a boat involving no symmetry in the repetition.

Molecules of arbitrary shape build three-dimensional networks in crystals. It helps to understand their structure if one imagines a two-dimensional model network in which the basic motif of arbitrary shape covers the entire surface, by repetition in two directions, without gaps or overlaps. The drawing in the figure shows an example constructing such a pattern.

The sides of the square contain halves of the fly, butterfly, falcon, and bat.

These sides serve as mirrors generating the two-dimensional pattern.

In crystal structures a large variety of networks occur, which are characterized by a variety of the symmetry relationship between the constituting molecules. Different symmetry relationships are depicted here in two of Escher's periodic drawings. The basic motif consists of one fish and one boat, and the pattern is created by their mere repetition.

In the next example, the basic motif is indicated by the highlighted square in which there are halves of one fly, one butterfly, one falcon, and one bat. The pattern is created by producing the mirror images of the basic motif. Whereas there is no symmetry involved in the repetition in the pattern of the fish/boat motif, here mirror symmetry is involved in creating the entire pattern.

MacGillavry worked closely with Escher to create a complete set of periodic drawings for aiding instruction in crystallography. Generations of students have benefited from this symbiotic interaction of the scientist and the artist.

Ágnes Csanády and Quasicrystals

For centuries, crystals used to be defined as solid-state matter with *regular* and *periodic* arrangements of their building elements. "Regular" meant a simple rule according to which its structural pattern could be generated or described. "Periodic" meant that the repetition of the building elements was virtually infinite. In contrast, amorphous matter has no regularity or periodicity in its internal structure. Examples for the former are a piece of diamond and a piece of graphite. An example of the latter is a piece of glass.

Dan Shechtman's discovery of the quasicrystals shattered this simple characterization. The internal structure of quasicrystals is *regular* but *nonperiodic.* It lacks translational symmetry, but it still fills all available space. The first quasicrystals were produced from aluminum-manganese alloys in 1982. It was difficult to accept the existence of quasicrystals because the definition of what a crystal was had become dogma in crystallography. When Shechtman published his discovery in 1984, an avalanche of publications followed. However, the official "seal of acceptance" came with Shechtman's Nobel Prize in Chemistry in 2011.

Long before this festive event, there appeared many papers with photographs reporting studies of quasicrystals. The materials scientist Ágnes Csanády and her colleagues in Budapest have produced the most beautiful quasicrystals at the development center of the Hungarian Aluminum Industry.

Ágnes Csanády's electron micrograph of flowerlike quasicrystals. The length of the horizontal bar corresponds to 1μm (equal to 0.0010 millimeters or 0.00004 inches). Courtesy of Ágnes Csanády.

Ágnes Csanády and Dan Shechtman, 1995, in Balatonfüred, Hungary. Photograph by Istvan Hargittai.

Israeli postage stamp commemorating the International Year of Crystallography, 2013.

In May 1995, there was an international conference on quasicrystals in a resort at Lake Balaton, Balatonfüred, Hungary. There, Csanády demonstrated her flowerlike quasicrystals. All, including Shechtman, were delighted. Csanády used the samples of the solid-state physicist Hans Ude Nissen of the ETH Zurich to produce her beautiful electron micrographs.

The Israeli Post Office honored the International Year of Crystallography (2014) in conjunction with Shechtman's Nobel Prize with a beautiful stamp, issued toward the end of 2013. It displays Csanády's quasicrystals and Shechtman's electron diffraction pattern of 1982, which led him to the existence of regular and nonperiodic structures in solid matter. For the representation of quasicrystals, the Israeli Post Office selected the most beautiful image that has ever been recorded of such structures.

5
Chemists and Biochemists

Chemistry and biochemistry are areas of science (as is crystallography) where circumstances have been relatively favorable for women joining the ranks of their male colleagues.

Gerty Cori

Gerty Cori. Source: Wikipedia.

According to tradition, one of the new laureates of each category gives a two-minute speech at the Nobel banquet. In 1947, for the Physiology or Medicine Prize, Carl Cori spoke. He expressed his satisfaction that his wife, Gerty, was also included in the prize. The Coris began their collaboration about 30 years earlier, at the time of their medical studies at the University of Prague, and it continued throughout their careers. He also stressed that their efforts were complementary, and they achieved much more in this combination than they would have, had each of them worked separately.

Gerty Cori (née Radnitz, 1896–1957) was born into a Jewish family in Prague, at the time in the Austro-Hungarian Empire, today, the capital of the Czech Republic. She studied at a private girls' school. When she decided to become a physician, she had to put in extra study hours to catch up in the science subjects. She enrolled at the medical school of the German University of Prague and became an MD in 1920. She met Carl Cori, a fellow student; both had a keen interest in biochemistry. They did research together and published a joint paper. They married in 1920, following graduation, and both worked in Austrian hospitals until Carl received an invitation from the New York State Institute for the study of malignant diseases in Buffalo.

Amid rising anti-Semitism in Central Europe, the Coris decided to emigrate. It took about half a year for Gerty to get an assistant pathologist's job in the same institute in Buffalo, after they arrived in 1922. They both appreciated that besides doing their work at the Institute, they were free to do research of their own interest, which eventually led to their Nobel Prize. During the nine years they spent in Buffalo they published extensively, mostly together, and they alternated the role of first author in their papers. Mildred Cohn, a distinguished scientist and one of their many students, told me, "They were remarkable. He would start a sentence and she would finish it. They were completely complementary. In personality, they were very different. He was aloof and she was vivacious, she was outgoing and he was not, although he was very insightful about people."[1] As the Coris became better and better known within the scientific community, Carl received attractive job offers, but none of them included the possibility of a job for Gerty; in fact, quite the contrary. The University of Rochester, for example, stipulated that he would get the job only if he stopped collaborating with his wife. When he

[1] Magdolna Hargittai, "Mildred Cohn," in *Candid Science*, vol. 3: *More Conversations with Famous Chemists*, ed. Istvan Hargittai and Magdolna Hargittai (London: Imperial College Press, 2003),258.

declined, they told Gerty that it is un-American for a man to work with his wife and that she should understand that by insisting on this, she was ruining his career.

Eventually they received an offer from Washington University in St. Louis, Missouri. He became full professor and chair of the Pharmacology Department and she received a position as research associate. In 1946, after nearly 15 years in this role, just a year before their Nobel Prize, Gerty became a full professor and Carl was made chair of the new Biochemistry Department. During their time at Washington University, they continued working together and producing significant results.

From the beginning, their main goal was to understand how energy is produced and transmitted in the human body. It was already known that during exercise the sugar that is stored in our body gets to the muscle, where it turns into energy. This process is called "carbohydrate metabolism." It was known that when carbohydrate metabolism does not work, this leads to diabetes. In 1923, Frederick Banting discovered that insulin could alleviate the symptoms of diabetes. This discovery, which was not a cure, only a treatment, further excited the Coris' interest in the chemical processes of carbohydrate metabolism.

After years of experiments, they came to significant conclusions. Sugar (glucose) is stored in our body in the form of a polysaccharide, called glycogen, in the cells of the liver and in the muscle. When we exercise, the glycogen in our muscles breaks down, and besides the needed sugar it produces lactate that diffuses from the muscle to the blood stream. The blood carries it to the liver, where it turns into glucose that is either stored there as glycogen or is transferred back to the muscle. The Coris called this order of events the "cycle of carbohydrates," but now it is simply known as the "Cori-cycle."

In the late 1930s, the Coris turned to enzymology. Enzymes are proteins that accelerate chemical reactions. The Coris recognized and isolated the enzyme called phosphorylase, which induces the breakdown of glycogen into glucose, together with other enzymes participating in these reactions. The Swedish scientist Hugo Theorell said in his presentation speech at the Nobel ceremonies in 1947, "For a chemist, synthesis is the definite proof of how a substance is built up. Professor and Doctor Cori have accomplished the astounding feat of synthesizing glycogen in a test tube with the help of a number of enzymes which they have prepared in a pure state and whose mode of action they have revealed. This synthesis would be impossible by methods of organic chemistry alone. . . . The Cori enzymes made this synthesis possible, because the enzymes favour certain modes of linkage."[2] The Cori laboratory in St. Louis became one of the most important centers in enzymology. It attracted talented researchers from all over the world. Beside themselves, six of their colleagues received Nobel Prizes.

In 1947, Gerty Cori was diagnosed with an incurable disease, a special form of anemia, and it took her life 10 years later. During her life, she had to face plenty of discrimination. Although the Coris worked and published together from the very first moments of their careers, for decades only he was recognized. He received many awards and distinctions alone; for example, he was elected to the National Academy of Sciences of the U.S.A. already in 1940, while Gerty was afforded the same recognition in 1948, only after her Nobel honor.

Alice Ball

Alice Ball (1892–1916) was the first woman and the first African American to receive a master's degree from the University of Hawaii (then the College of Hawaii). Then, she was the first woman and first African American professor of chemistry at the university. She was born in Seattle, Washington, into a well-to-do family. Her grandfather was a photographer and her parents followed in his profession, but her father was also a lawyer and a newspaper editor. Both her parents are listed as "white" on her birth certificate, probably to help her later position in society. Alice graduated from the Seattle High School in 1910 with top grades. She continued at the University of Washington in Seattle and earned a bachelor's degree in pharmaceutical chemistry in 1912 and another in pharmacy in 1914. Her advisor in pharmacy, Williams Dehn, and she co-authored a research paper in the prestigious *Journal of the American Chemical Society.*

She received offers of scholarships for graduate studies. One of them came from the University of California, Berkeley, but she chose Hawaii. For her master's thesis, she became involved in improving the treatment of leprosy patients. She joined the attending physician at the local hospital, Dr. Harry T. Hollmann, in his research on the chaulmoogra oil from the

[2] H. Theorell, "Presentation Speech," in *Nobel Lectures, Physiology or Medicine 1942–1962* (Amsterdam: Elsevier, 1964), 181.

Alice Ball, 1915. Source: Wikimedia.

seeds of the *Hydnocarpus wightianus* tree. Leprosy patients were treated with this oil in the form of an injection. However, this oily, viscous substance accumulated in clumps under the skin and caused the skin to appear dotted with blisters.

Ball developed a procedure to make the injected substance absorbable by the body while retaining its therapeutic values. She died before she could have published her innovation, and it proved easy to expropriate it. A high-ranking professor at the college published her results without giving her credit and benefited from the production and sales of the substance. A long uphill battle first by Hollmann and later by others at what is now the University of Hawaii resulted in the recognition of Ball's contribution. Archival materials proved her original work and achievements that we can now honor.[3]

[3] "Alice Ball," Wikipedia, accessed August 23, 2022, https://en.wikipedia.org/wiki/Alice_Ball.

Ida Noddack

Ida Noddack. Courtesy of William Jensen and the Oesper Collections in the History of Chemistry, University of Cincinnati.

In 1934, Enrico Fermi and his associates at the University of Rome bombarded uranium (atomic number 92) with neutrons and supposed that the products contained two new elements heavier than uranium. In the same year, the German analytical chemist Ida Noddack (née Tacke, 1896–1978) suggested an alternative interpretation, according to which the neutron bombardment broke up the uranium nucleus yielding atoms of two elements lighter than uranium. Although she published and publicized her ideas, they were ignored. This was unfair to her personally, of course, but from a world-historical perspective, it was perhaps fortunate that nuclear fission was not discovered five years earlier than it officially was. In December 1938, Fermi received the Nobel Prize in Physics, in part for the discovery of the new heavy elements. In the same month, Otto Hahn and Fritz Strassmann in Berlin demonstrated the presence of barium, a known element lighter than uranium, among the products of neutron bombardment of uranium. Soon, Lise Meitner (see chapter 3) and Otto Robert Frisch interpreted the experiment as nuclear fission, thus vindicating Noddack's earlier supposition.

Ida Tacke was born in Lackhausen (now Wesel) in northwest Germany. She studied chemistry and metallurgy at the Technical University in Berlin-Charlottenburg as one of its first female students. In 1925, she moved to the Physikalische Technische Reichsanstalt (Imperial Institute for Physical Technology) in Berlin. There she worked with Walter Noddack, who was looking for missing elements in the periodic table of the elements. Tacke and Noddack married in 1926.

In 1925, together with Otto Berg, they published the discovery of two new elements and named them masurium (atomic number 43) and rhenium (atomic number 75). The name "rhenium" referred to the River Rhine and the region where Ida was born. The name "masurium" referred to the region Masuria, then in Prussia, today in Poland. The naming of the two elements was interpreted as an expression of German nationalism. The discovery of rhenium was accepted by the scientific community, but not of masurium. Nobody could reproduce its extraction from the ore from which the Noddacks claimed to have extracted it. Decades later, when a new element called technetium was discovered, it turned out to be identical to what the Noddacks had named masurium. This was a belated vindication of the Noddacks' claim.

Ida Noddack did not have a formal position. In the late 1920s and early 1930s, at the time of the Great Depression, women in Germany were not encouraged to work. Even those who did were often forced out of their jobs when they married in order to make their places available for men. Ida continued her research in her husband's laboratory, unpaid but as a full-fledged researcher. In 1935, the Noddacks moved to the University of Freiburg, where they worked at the Institute of Physical Chemistry.

In 1941, after Germany occupied the Alsace region in northeastern France, Walter Noddack was appointed professor and director of the Institute of Physical Chemistry at the newly founded Reichsuniversität Strassburg (Strasbourg Imperial University). The French Université de Strasbourg had gone into exile and stayed there until the liberation of France. At the German university, Ida had a paid position. The Noddacks received generous assistance in equipment and other means for their research. In spite of their much improved situation, however, they did not publish anything during their Strasbourg period, and it ended with the end of Nazi Germany.

It has been much discussed in the literature whether or not the Noddacks were Nazis or Nazi sympathizers. At Reichsuniversität Strassburg, 80 percent of the full professors in chemistry were members of the Nazi Party, but neither of the Noddacks were.[4] The fact, however, that they were offered positions there indicates that they were sufficiently trusted by the Nazi regime; in any case, that they accepted these positions bears witness to their willingness to benefit from the Nazi reign.

In November 1944, when the Allied Forces were approaching Strasbourg, the Noddacks packed up their equipment and sent it to Germany. There, eventually, they received permission from the Allies' military government to continue their research. Possessing the equipment helped them find a position at the end of 1946 at the Philosophisch-Theologische Hochschule (Philosophical and Theological College) in Bamberg. Walter founded the private Institute of Geochemistry with the instruments from Strasbourg. Ida continued her research in the Institute, again unpaid. The Institute was nationalized in 1956 and Walter became its director, a position he held until his death in 1960. Ida continued working there until 1968, when she moved to a retirement community near Bonn.

Irène Joliot-Curie

Irène Joliot-Curie (1897–1956) was the child of the pioneers of radioactivity, Marie and Pierre Curie, and herself became a pioneer in this field. She was still very small when her parents received the Nobel Prize. With both her parents working in their laboratories, she was brought up by her physician paternal grandfather, Eugene Curie. He taught her botany and natural history and inculcated in her the significance of democratic and socialist ideals. She adhered to these ideals for life.

She was a student for her bachelor's degree when World War I broke out. She learned nursing and radiology to be able to work together with her mother. They built X-ray machines for diagnosing the wounded in the war.

[4] B. Van Tiggelen and A. Lykknes, "Ida and Walter Noddack through Better and Worse: An Arbeitsgemeinschaft in Chemistry," in *For Better or Worse: Collaborative Couples in the Sciences*, ed. A. Lykknes et al., Science Networks, Historical Studies 44 (Basel: Springer, 2012), 113.

Irène and Marie Curie, 1925. Source: Creative Commons.

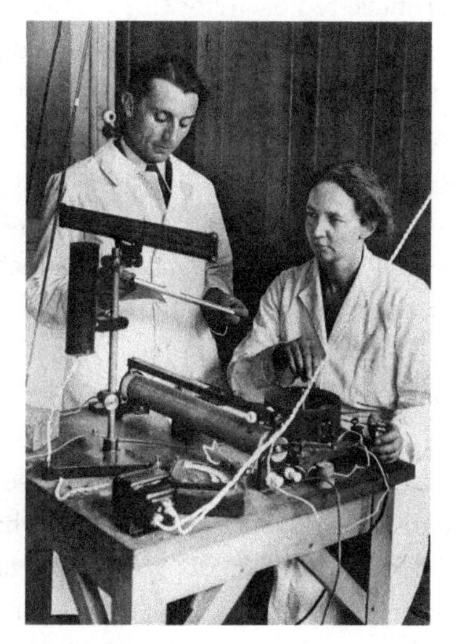

Frédéric and Irène Joliot-Curie, 1940s. Source: Frédéric Joliot-Curie tudománya a boldogságért - Cultura.hu.

Irène was responsible for installing and using the equipment in a radiological facility in Belgium. Beyond creating and operating these machines, she had to overcome serious resistance among people in general and especially among older physicians. They found the X-ray machines alien and frightening. She taught doctors how to locate shell fragments and shrapnel parts in the X-ray pictures, and she spent much of her time in front of the radiography machines. It is likely that in this way she received large amounts of radiation that would eventually kill her.

Irène always wanted to become a scientist and work in the field of her parents. After the war, she started at the Radium Institute at the Sorbonne as her mother's assistant. She worked on her doctoral thesis, which she submitted in 1925. A thousand people attended the public defense of her dissertation.

When a new assistant, Frédéric Joliot, came to work in their laboratory, Irène was charged with teaching him the laboratory techniques. They went on to work and publish together, and in 1927 they married. Eventually, he added the Curie name to his surname. One of their joint discoveries was the method of calculating the accurate mass of the neutron. Their most important result was the discovery of artificial radioactivity. In 1934, when they put polonium next to aluminum, they observed radioactive emission from aluminum. Even when they took away the polonium, the emission continued. An unstable isotope of phosphorus was produced in the process, which is not found in nature. This was an alchemist's dream: turning one element into another, in this case, turning aluminum into phosphorus. At the time, there was a growing use of radioactive materials in medicine, and their method made the production of new radioactive materials possible on the spot. The discovery had other applications as well, in nuclear technologies. Nobel recognition followed quickly on the heels of their discovery. In 1935, Irène and Frédéric jointly received the Chemistry Prize for the synthesis of new radioactive elements.

In their rich research career, they sometimes missed what would have been other milestone discoveries. Such was the case when they did not recognize heretofore unknown fundamental particles, the neutron on one occasion, and the positron on another. James Chadwick discovered the neutron and received the Nobel Prize for it in 1935, and C. D. Anderson discovered the positron and received the Nobel Prize for doing so in 1936. There are indications that in 1938 Irène might have discovered nuclear fission but did not recognize it. On this occasion, she worked alone.

After World War II, the Joliot-Curies were among the scientists who created the first French nuclear reactor and took leading positions at the

French Atomic Energy Commission. Their most productive scientific period was when they worked together in research. Initially, the situation in their joint work was the reverse of the usual: Irène was the leader and the authority and Frédéric had to prove that he was on equal footing with her. Both of them had public positions. Frédéric was a dedicated communist and an uncritical friend of the Soviet Union.

Irène and Frédéric had two children, Hélène Joliot and Pierre Joliot. Both became scientists, she a nuclear physicist and he a biophysicist. Hélène married Michel Langevin, the grandson of the great French physicist Paul Langevin. Three Curies have become members of the French Academy of Sciences: Pierre Curie, Frédéric Joliot-Curie, and Pierre Joliot. None of the women has been elected, not Marie Curie, not Iréne, and not Hélène Langevin-Joliot. Both Marie and Irène had been proposed but were refused. The first woman elected to the French Science Academy (in 1962) was a student of Marie Curie, Marguerite Catherine Perey (1909–1975), for her 1939 discovery of the element francium.

Klavdia V. Topchieva

Klavdia V. Topchieva (1911–1984) graduated from the Lomonosov Moscow State University (MSU), where she later served as professor of chemistry and for a time as dean of the Chemistry Department. She was a physical chemist and her main field was catalysis. She was born near the town of Morozovsk in the region of Rostov-on-Don in southern Russia, the only daughter in a family of factory workers. Each of her four brothers became highly successful in Soviet society, occupying important positions in government, the Academy of Sciences, and the defense establishment. Having ambitions and successful brothers put some pressure on Topchieva, which she withstood brilliantly. In everything she did she was always aiming at the top. Not only did she succeed in her goals, but she remained popular among her peers.

She enrolled in the Chemistry Department at MSU in 1929. Parallel to her undergraduate studies and throughout her graduate school and fledgling years as a scientist, she was seriously engaged in playing volleyball. She was on teams winning tournaments and championships and was one of the first volleyball players to be awarded the title of Master of Sport of the Soviet Union (1938). This was the same year she earned the PhD-equivalent degree of Candidate of Science.

The memorial of Klavdia V. Topchieva. Photograph by M. Hargittai.

She started her career in research in the Laboratory of Kinetics and Catalysis at MSU. Kinetics is concerned with the speed of chemical reactions; catalysis entails the use of special materials to speed up chemical reactions and allow them to proceed under milder conditions. Both have enormous significance in industry.

In 1953, Topchieva defended her dissertation for the higher doctorate, DSc, which is a prerequisite for a professorial appointment in the Soviet/Russian system. The following year, she was appointed professor, and in short order became the dean of the Department of Chemistry (1956–1960), the first woman in that position. Her deanship was short, but she left a lasting impact on the life and development of the department. This department is a very large institution, and she proved to be an efficient and accessible leader. She remained an authority among her peers for the rest of her life.

As a scientist, she created a substantial school within the department with the principal direction of research on zeolites. Zeolites are porous materials whose main building elements are aluminum, silicon, and oxygen. They occur in nature as minerals and can also be produced in the laboratory. Their significance is in their application as catalysts, hence their importance for practical applications on an industrial scale, even in biotechnology. Topchieva's distinguished career at MSU, from undergraduate studies to retirement, lasted 55 years.

Mildred Cohn

Mildred Cohn, 2002. Photograph by M. Hargittai.

"Here is an example of a woman who had tremendous ability, but who was in secondary positions in academia practically until the time when she was elected to the National Academy of Sciences. Mildred continued to make important contributions to enzymology and oxygen-18 measurements that influenced my research over the years."[5] So said the Nobel laureate Paul D. Boyer in 1999, reflecting on Mildred Cohn (1913–2009). The use of tracers, such as oxygen-18,[6] made it possible to understand the mechanism of certain crucial reactions in living organisms at the molecular level. Cohn first introduced Boyer to the technique via correspondence. Boyer contacted her after he had read one of her papers. They met in person much later.

Cohn was born in New York City into a Jewish-Russian immigrant family. She had an outstanding high school teacher who attracted her to chemistry. She studied at Hunter College, where her favorite subject was physics, but the school did not yet offer it as a major. So Cohn graduated in chemistry with physics as her minor. She applied to 20 graduate schools, but there was no offer for her, so she used her savings to attend Columbia University, where at that time only men could receive a teaching assistantship. She ran out of money after the first year and took a job, but still managed to continue her studies. In 1934, she signed up to be Harold Urey's doctoral student just months before the announcement of Urey's Nobel Prize in Chemistry for the discovery of deuterium. She began working with the oxygen isotope oxygen-18 in Urey's lab, and stayed with him until 1937. She made an interesting observation about the variation of the number of women graduate students at the time and subsequently:

> People are under the impression that there were no women around in those days. The statistics show that there was a larger fraction of female graduate students then than there were in the 1950s and 1960s. There was a real drop in the 1950s and 1960s in the percentage of women PhDs awarded in this country in science. My interpretation is that this drop occurred because the men came back from World War II and the women were displaced. During the war years, women took up a lot of non-traditional jobs. But afterwards, to get the women out, they were told that women should

[5] Istvan Hargittai, "Paul Boyer," in I. Hargittai and M. Hargittai, *Candid Science*, vol. 3, 274.
[6] Oxygen-18 is a rare isotope of oxygen whose common occurrence is oxygen-16.

> get married and have children, and if they did not bring up their children themselves until they were at least five years old, the children would become monsters and so on. That had an effect. It took a while before women returned to attend graduate school again, not until the women's liberation movement started.[7]

Cohn had superb mentors even after she graduated from Urey's tutelage. First, she worked in Vincent du Vigneaud's group at Cornell University, then in Carl and Gerty Cori's laboratory at Washington University in St. Louis. Urey, du Vigneaud, and the Coris were all future Nobel laureates. It took Cohn two decades before she was finally given her own professorial appointment, at the University of Pennsylvania. She was eventually appointed to the prestigious position of Benjamin Rush Professor of Biochemistry and Biophysics. In 1971, she was elected to membership in the National Academy of Sciences. One of her many distinctions was the National Medal of Science in 1982, presented to her by President Ronald Reagan.

Cohn's research activities focused on the utilization of isotopes, about which she became a great authority. Her favorite physical technique was mass spectrometry, and her research resulted in seminal results both in conceptual and methodological aspects. She contributed to the knowledge about how enzymes work by applying nuclear magnetic resonance spectroscopy based on the phosphorus-31 isotope. She studied the enzyme reactions of adenosine triphosphate (ATP), and showed that the extent of muscle disease can be followed according to the variations in ATP concentration. She also determined the concentration of magnesium in the human brain.

Cohn's husband, the theoretical physicist Henry Primakoff, was an immigrant from Russia. The two met in 1934 at Columbia University. Her venues of research were largely determined by her husband's appointments, but like so many other women at the time, she was barred from employment at these institutions because of anti-nepotism rules. Primakoff never took a job without making sure that his wife could do research there as well. Her career was important to him, but this was the extent of his support. As far as participating in household chores, Cohn put it this way: "[He] was very European in this regard. Whenever I asked him to do something, he said,

[7] M. Hargittai, "Mildred Cohn," 257.

'Hire somebody.' He never participated in housework, but often played with the children and invented stories for them. . . . I was the practical one in the family. If a child broke a toy, I was the one who repaired it. He never did anything practical, not only traditional women things, he never fixed the car, he was not interested in the garden, and so on. He was very cerebral."[8] That Cohn could not have a faculty position had one advantage: she could devote herself exclusively to her projects. She was eager to perform well in the lab, and for 30 years they had an excellent woman help to take care of their three children. Being a working mother was unusual at the time. Their oldest child, Nina, was the only one in her school whose mother had a job, and she complained about this. In college, Nina majored in psychology and wrote a paper about the effects on children of having a working mother versus a non-working mother. She concluded that, in the end, there was no appreciable difference. All three children earned their PhDs; the two daughters became psychotherapists and the son a biochemistry professor.

To my question about her advice for a young woman who would like to do science and have a family too, Cohn replied, "The first thing I would suggest is, marry the right man. That's the most important thing. You have to have a husband who is fully supportive. That he does more than paying lip service to equality. My husband was really a feminist. He liked women and respected them. My second advice is that whatever decision they make they shouldn't feel guilty."[9]

Gertrude B. Elion

Many quotes covered the walls in Gertrude Elion's office at Glaxo Wellcome in Research Triangle Park, North Carolina, when my husband and I visited her in 1986. One of her fans wrote:

> In 1984, our 5-years-old daughter was diagnosed with Acute Lymphocytic Leukemia and was put on a chemotherapy protocol including, among other drugs, 6-mercaptopurine. Our family went through many difficult times, both physically and emotionally. We are thrilled to tell you that our daughter has been in remission for over 5 years, with no relapses, and has

[8] M. Hargittai, "Mildred Cohn," 262.
[9] M. Hargittai, "Mildred Cohn," 263.

Gertrude B. Elion in her office in the Elion-Hitchings Building at Glaxo Wellcome in Research Triangle Park, North Carolina, 1996. Photograph by M. Hargittai.

> been completely off medication for 2 years, 3 months, and 24 days! She will celebrate her 11th birthday in May and is a delight to her parents. When we see the adoration in the media of overpaid sports figures and entertainment figures with inflated egos, we can only think how much more you and your colleagues have contributed to society, with little or no recognition. You are truly a hero.[10]

Gertrude Elion (1918-1999) was born in New York City into a scholarly family. Her parents lost everything during the Great Depression, but they knew

[10] Istvan Hargittai, "Gertrude B. Elion," in *Candid Science: Conversations with Famous Chemists*, ed. Magdolna Hargittai (London: Imperial College Press, 2000), 71.

that their children had to get a good education because that was the only way to a better life. From early childhood she loved to read, and her favorite book was Paul de Kruif's *Microbe Hunters*. The book has been in print since it was first published in 1926 and has fascinated many young readers about scientists and careers in science.

Elion wanted to become a chemist so that she could find the cure for cancer. She had a painful personal motivation for her aspirations: her grandfather died of cancer just before she entered college. Later, her determination to find cures for sick people further strengthened when her fiancé died. He suffered from a disease that could have been cured by penicillin, which became available shortly after he died. Elion spectacularly succeeded in fulfilling her goals in creating drugs to cure a variety of diseases, more than she could have imagined. In 1988 she received the Nobel Prize in Physiology or Medicine, together with James W. Black and George H. Hitchings, for establishing novel principles for drug treatment, called "rational drug design," which refers to the development of drugs based on the study of the structure and functions of the target molecules.

Her road to the highest possible recognition in science was not easy, however. Like so many others in her day, Elion attended Hunter College, graduating with highest honors. After college, she applied to 15 graduate schools, but none accepted; young Jewish women were not wanted. She could not finance herself, so she accepted teaching positions to pay for her education. She enrolled at New York University and did her research on nights and weekends. She received her master's degree in 1941, but to become a researcher, her dream and goal, was difficult without a PhD. A break came when George Hitchings interviewed her at Burroughs Wellcome Pharmaceutical Company. He was not interested in whether or not she had a doctorate, only whether she could do the job. She had good grades and knew German; that was important because at that time much of the scientific literature was published in German. Hitchings brought her on as his assistant.

Hitchings followed the philosophy of the company's founder, Henry Wellcome, who supported researchers if they had good ideas. Hitchings never told Elion what to do or not to do. He gave her the opportunity to do as much as she could. His research aimed to block the survival of bacteria, parasites, and tumor cells by replacing certain building blocks in their deoxyribonucleic acid (DNA), and he wanted to find molecules whose building blocks were similar. Anything that had to multiply in order to survive had to make DNA. This was revolutionary not only because it was different from the usual approach in drug research, but also because it focused on DNA, which not many researchers were interested in yet.

Elion prepared a variety of purine derivatives. Purine is a simple organic molecule; two of the four bases of DNA are purine derivatives (the other two are pyrimidine derivatives). Purine itself consists of a six-member ring fused together with a five-member ring. By giving the patients various derivatives of this molecule, Elion hoped to fool the bacteria, parasites, or tumor cells. She did countless organic syntheses, and she also learned the related fields so that she herself could test the biological activities of the new substances. The hard work bore fruit; during the three decades of their joint work, Hitchings and Elion developed a number of drugs, all of them derivatives of purine. Moreover, their drugs developed for certain treatments proved efficient in treating other conditions as well. First they were surprised by such coincidences, but eventually they understood that since they were targeting the DNA of the culprits of various diseases and disorders, the cure they were offering had a substantially broader spectrum than originally envisioned.

There then occurred yet another use for one of the substances they applied, 6-mercaptopurine, in muting the body's immune response following organ transplantation. The immune system of the organism provides protection against unwanted invaders. However, a transplanted organ may also be considered an unwanted invader, and in such a case the immune system needs to be muted. A simple derivative of 6-mercaptopurine, the drug Imuran, was mentioned with great appreciation in the surgeon Joseph E. Murray's Nobel lecture in 1990. He reported about the successful organ transplants in humans and referred to his close cooperation with Hitchings and Elion. Before Imuran became available, kidneys could be transplanted only from one identical twin into the other. Imuran made it possible to perform this procedure for many hundreds of thousands of nonrelatives without such limitation. Before the successful applications in humans, tests were carried out on dogs.

At the age of 70, Elion finally found herself in the limelight when she received the Nobel Prize. At first, she thought the Prize would not change her life, but it did. All of a sudden she was in great demand to sit on committees and advisory boards and give talks. She found it fortunate that the award came rather late in her life so that these sorts of demands did not interfere with her creative period. Elion died in 1999 at the age of 81. She never married. She remained active to her last day, ever full of plans.

From left to right: the dogs Tweedledum and Tweedledee, unknown, Roy Calne, the dog Titus, Gertrude Elion, the dog Lollypop, George Hitchings, Donald Searle, E. B. Hager, and Joseph Murray. Courtesy of Gertrude B. Elion and Katherine T. Bendo.

Maxine F. Singer

Thanks to "her outstanding scientific accomplishments and her deep concern for the societal responsibility of the scientist," in 1992 Maxine Singer was awarded the National Medal of Science, presented by President George H. W. Bush.[11] This quotation summarizes her achievements; her multifaceted accomplishments as a policymaker, a science organizer, and a scientist are exceptional.

Maxine F. Singer (née Frank, b. 1931) was born in New York City. Her father was a lawyer and her mother a homemaker until World War II, when she went to work, liked it, and continued after the war's end. Maxine's interest in science started in high school, due to the influence of her science teachers.

[11] National Science Foundation, "The President's National Medal of Science: Recipient Details," accessed April 13, 2014, http://www.nsf.gov/od/nms/recip_details.jsp?recip_id=327.

Maxine F. Singer, 2000. Photograph by M. Hargittai.

She graduated in 1952 from Swarthmore College, a small coeducational school in Pennsylvania. The years she spent there had the greatest impact on her development as an independent scientist:

> Just by luck in my year the outstanding science students were women. There were six of us. We were close friends, we lived in the same dormitory and to a very large extent we educated one another. I tend to think that if as an undergraduate, I hadn't been in that group, I might not have had the will and the ambition to keep on going in science. I could easily have gone to medical school, maybe. It was the fact that I had that very strong group of friends for four years that made a tremendous difference. Thus, I think that

my attitudes about science probably come much more from my undergraduate education than from my graduate education.[12]

Singer received her PhD in biochemistry from Yale University. On the suggestion of her doctoral advisor, Joseph Fruton, she went into DNA research and joined the Laboratory of Biochemistry at the National Institutes of Health (NIH). She was involved with molecular biology and biochemistry, and for some time she worked alongside the 1968 Nobel laureate Marshall Nirenberg to decipher the genetic code. Assisting Nirenberg was her most important contribution to research.

After 17 years at the NIH, Singer moved to the NIH National Cancer Institute, where she held leading positions. Meanwhile, she became more and more involved with ethical issues related to genetic research. For eight years, beginning in 1980, she was the director of the Laboratory of Biochemistry at the National Cancer Institute. In 1988, she became president of the Carnegie Institution in Washington, D.C., a large private research organization in three fields: astronomy, biology, and earth sciences. As president, she founded First Light, a Saturday science school for children. This was the start of a new program, the Carnegie Academy for Science Education, whose aim was to reach out to students and teachers in D.C. area schools to encourage their interest in science. She also founded a new department for research in global ecology, before retiring from her position in 2002. Singer has been well recognized for her accomplishments: her distinctions include memberships in the National Academy of Sciences of the U.S.A. (1979) and the Pontifical Academy of Sciences (1986); she also received the National Medal of Science (1992) and the Public Welfare Medal from the National Academy of Sciences (2007).

Going back a few decades, her role in initiating discussions about the possible consequences of genetic engineering deserves special mention. It was at a meeting on nucleic acids in 1973 where the potentially harmful effects of the recently developed recombinant DNA technologies came up. "Recombinant DNA technologies" refers to editing—the possibility of cutting and then joining together—the DNA of different species. This raised serious concerns about the dangers of creating new and potentially disruptive organisms. In 1975, together with Paul Berg and other leading scientists in

[12] Magdolna Hargittai, conversation with Maxine Singer, May 16, 2000, Washington, D.C. All unreferenced quotations are from this conversation.

the field, Singer organized the famous Asilomar Conference to discuss the potential hazards of this new technology, later calling it "a defining moment for a generation, an unforgettable experience, a milestone in the history of science and society."[13] Among the participants were leading researchers in the field, as well as physicians and lawyers, to discuss whether the dangers were real and, if so, what should be done to prevent or mitigate them. Although most of the scientists thought that there was no real danger, they understood that the stakes were too high to ignore. They prepared guidelines for future research—imposing self-restrictions on recombinant DNA research—and they agreed that the restrictions would be lifted gradually as new results prove the research safe. The Asilomar Conference was exemplary because the scientists understood the hazards and their responsibilities and acted upon them rather than waiting for legislative action.

During the following years, Singer remained active in keeping the public informed about genetic engineering, the Human Genome Project, and other scientific issues. She testified before the U.S. Congress and served as an advisor on legislative committees dealing with these topics. Could it be that the caution in dealing with recombinant DNA was an overreaction? She did not think so. In our age, science has tremendous impact on the life of society, and it uses enormous amounts of public resources. Because of this, the task of scientists is not only to make discoveries but also to inform the public. The question is whether the scientific community does this adequately. This was a relevant and urgent issue at the dawn of genetic engineering.

Singer has been a prolific writer. She has about 100 scientific publications and has co-authored several books with the Nobel laureate Paul Berg. Singer was lucky in never having to experience gender discrimination, neither as a student nor later as a researcher. Her only noteworthy experience in this respect came early on, when she was hiring post-docs and found there was scant interest in going to work in her lab, that is, in working for a woman. This problem gradually disappeared.

Maxine Singer has had a spectacular career as a scientist as well as a science administrator and policymaker—and all that while having a happy family life with four children. This was her greatest challenge: raising four kids and having them turn out wonderfully.

[13] M. Barinaga, "'Asilomar Revisited' Lessons for Today?," *Science* 287, no. 5458 (2000): 1584–1585.

Elena G. Galpern

Elena G. Galpern. Courtesy of Elena G. Galpern.

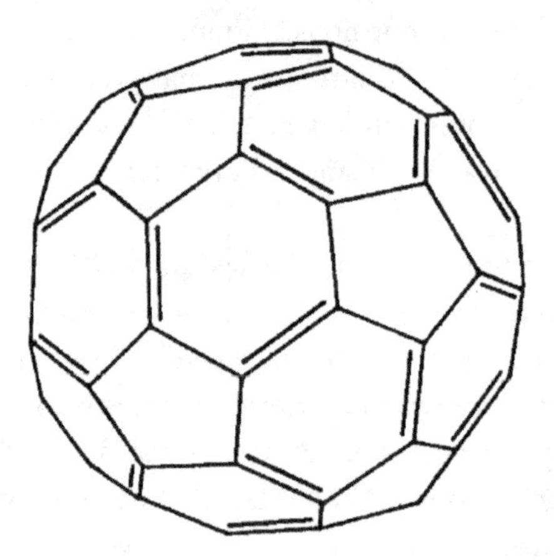

A model of the C60 all-carbon molecule, buckminsterfullerene.

In 1985, the discovery of a new form of carbon, besides graphite and diamond, made the news. The molecule of this new form looked like a soccer ball, consisting of 60 carbon atoms at the apexes of a truncated icosahedral shape, and was called "the most beautiful molecule." The three principal discoverers, Robert F. Curl, Harold W. Kroto, and Richard E. Smalley, received the Chemistry Nobel Prize in 1996.

There was a lot of excitement over the molecule, and as time passed, we learned that more than a decade prior to the celebrated discovery, there were two publications in which such a C_{60} molecule had been predicted. Unfortunately, both of these publications appeared in not very accessible journals, one of them in Japan and the other in the Soviet Union, although the Soviet journal existed in full English translation in the West. The Japanese paper, by Eiji Osawa, reported a *suggestion* of the truncated icosahedral shape of a C_{60} molecule. The Russian paper came to a similar conclusion based on quantum chemical computations performed by a junior researcher, Elena G. Galpern (b. 1935).

In the early 1970s, Elena was working toward her PhD-equivalent degree at the Laboratory of Quantum Chemistry of the Institute of Element-Organic Compounds (INEOS) of the Soviet Academy of Sciences; she was a research associate of Dmitrii A. Bochvar (1903–1990), who established the laboratory years earlier and determined the general direction of its work. The director of INEOS, Aleksandr N. Nesmeyanov, was one of the top scientists and science administrators in the Soviet Union, high in the hierarchy, but he was also an original thinker in chemistry. At one time, he had the idea to produce cage-like molecules consisting of carbon atoms, where the cage would accommodate an atom of a different element, or even small groups of atoms. Nesmeyanov envisioned a plethora of uses for such substances. The first task was to find the carbon cages that could house such "hetero" atoms. The emerging technique of quantum chemical calculations appeared to be suitable to determine the feasibility of such structures, and the project seemed appropriate for Galpern's dissertational research.

The carbon cages were quite large systems for the computations then available, hence they had to start with the smallest cages and gradually advance toward the larger ones. This is how Galpern arrived at the systems consisting of 60 carbon atoms. She had to find a shape that would be sufficiently stable, and in the course of her computations, she had already tested many shapes. There were yet so many other possibilities that hoping to test all seemed futile. One day, her senior colleague, Ivan Stankevich, who had just returned from a soccer match, suggested testing the shape of the soccer ball. It was only a few years before that they started producing soccer balls by sewing together pentagonal and hexagonal patches. The ball is, of course, a sphere whose polyhedral analog is a truncated icosahedron consisting of pentagonal and hexagonal sides and possessing 60 apexes. Galpern's computations showed that such a shape of 60 carbon atoms would be stable indeed. In soccer, of course, the 22 players of the two teams kick that shape for 90 minutes and it withstands the abuse—it has to be sturdy. The logic was that an all-carbon molecule taking this shape would also be stable.

When Galpern completed her computations, she prepared a manuscript for publication, and her boss, Bochvar, decided to bring it out in the most prestigious Soviet periodical, *Dokladi Akademii nauk SSSR* (Proceedings of the Soviet Academy of Sciences).[14] It duly appeared in 1973, and although an English translation of the journal made the article accessible for non-Russian-readers, the discovery remained unnoticed both in the Soviet Union and in the rest of the world.

In 1985, Harold Kroto of Sussex University in England and Richard Smalley and Robert Curl of Rice University in Houston, Texas, along with their students, observed the stable C_{60} in an experiment at Rice. When they suggested the truncated icosahedral shape for it, this was a turning point in the story. This beautiful molecule was no longer just in Osawa's dreams and Galpern's computer, but was observed in an experiment. Soon, combing the literature, the researchers spotted Osawa's suggestion and Galpern's computations. Osawa and Galpern received considerable attention as a result, though far less than Kroto, Smalley, and Curl.

Galpern was bewildered by the sudden publicity, even if it was short-lived. She had had a quiet career, nothing spectacular; all this time working at INEOS, she had not realized that she had published her most important results in the course of her graduate studies. One could speculate how Galpern's career might have turned out had she or her professor realized the importance of their C_{60} molecule, what might have happened had somebody noticed her report and tried to produce the C_{60} molecule and thus recognized the importance of her work years earlier. It seems that Nesmeyanov may have lost interest in the carbon cages and directed the attention of his associates elsewhere. When fame finally reached Galpern, she did not dwell much on what might have happened; she was happy to see so much interest in her early work, even though it was belated.

[14] D. A. Bochvar and E. G. Galpern, *Dokladi Akademii nauk SSSR* 209 (1973): 610–612.

Joan A. Steitz

James D. Watson and Joan A. Steitz, 2003, in Cambridge, UK, during the celebrations of the 50th anniversary of the discovery of the double-helix structure of DNA. Photograph by M. Hargittai.

Joan A. Steitz (née Argetsinger, b. 1941) is a pioneer in molecular biology among women researchers and has become a world authority in ribonucleic acid (RNA) biology. Her discoveries have earned her awards and distinctions. She was a latecomer in recognizing gender discrimination in science and has fought for more equitable treatment ever since. She grew up in Minneapolis, Minnesota, attending a girls' high school, Northrop Collegiate School. She enrolled at Antioch College in Ohio and received her bachelor's degree in chemistry in 1963. She did an internship in Alex Rich's laboratory of X-ray crystallography at MIT, well known for its structure determination of biological molecules. This experience made her interested in molecular biology.

She enrolled at the new biochemistry and molecular biology program at Harvard University and was the first female graduate student in James D. Watson's laboratory. Watson was already a Nobel laureate and one of the leaders in the field. Steitz participated in the research of bacteriophage RNA, bacteriophages being the viruses infecting and replicating in bacteria. For her postdoctoral studies she went to Cambridge, UK, where she worked at the Laboratory of Molecular Biology of the British Medical Research Council. There, she gained research experience with such giants of molecular biology as Francis Crick and Sydney Brenner. Among her discoveries was the sequence of a virus RNA that encoded three proteins.

In 1970 she joined Yale University and embarked on her iconic investigation of the mechanism of the action of messenger RNA (mRNA) in protein production. She concluded this investigation in 1975 and continued her fundamental research in RNA biology. For example, she discovered how modifications of mRNA lead to the production of slightly different proteins performing slightly different functions in the organism. Her results in fundamental research started transforming into medical applications, when, for example, the task is to find a remedy when a patient's body produces antibodies against the patient's own nucleic acids or ribosomes.

Probably because she has always been an outstanding scientist working under privileged conditions, only when she had become chair of the Department of Molecular Biophysics and Biochemistry at Yale did she recognize and understand that even innocent-looking administrative measures might have a disproportionate impact on women. She concluded from her own experience, "If a woman is a star there aren't that many problems. If she is as good as the rest of the men, it's really pretty awful. A woman is expected to be twice as good for half as much [recognition]."[15]

Her discoveries have earned her the highest awards and distinctions. She was elected to the National Academy of Sciences of the U.S.A. in 1983 and received the National Medal of Science from President Ronald Reagan in 1986. The L'Oréal-UNESCO Award for Women in Science followed in 2001, then the Gairdner Foundation International Award in 2006. She won the Lasker-Koshland Award for Special Achievement in Medical Science in 2018, and the

[15] Joan A. Steitz | National Science and Technology Medals Foundation (nationalmedals.org)

Wolf Prize in Medicine in 2021. Her husband, Thomas A. Steitz (1940–2018), was a co-recipient of the Nobel Prize in Chemistry in 2009 for studies of the structure and function of the ribosome. They were both professors at Yale University and had one son.

Lynne E. Maquat

Lynne E. Maquat, 2021. Photograph by the University of Rochester. Courtesy of Lynne E. Maquat.

Lynne E. Maquat (b. 1952) has been a pioneer in at least two senses, first for her research on the biology of RNA in which she paid equal attention to understanding the function of RNA and the role of its changes in inherited and acquired diseases, and second in helping fledgling female scientists on their way to becoming confident and successful researchers. She did her undergraduate studies in biology at the University of Connecticut, Storrs, and earned her bachelor's degree in 1974. She continued in biochemistry at the University of Wisconsin at Madison. There were no female professors in the Department of Biology then (today there are 10 among 24 faculty members). The male professors, at least some of them, were apprehensive about having female graduate students; some even declared that women had no place in their science. Maquat and two other female graduate students ignored their hostility and met regularly to encourage each other to succeed. Apparently this worked. After finishing her PhD in 1979, she moved to the McArdle Laboratory for Cancer Research in Madison for her postdoctoral studies. Then she spent 18 years at the Roswell Park Cancer Institute in Buffalo, New York.

When she joined the University of Rochester, her career took off in a big way. She and her husband, Mark Spall, a technology consultant, have lived in Rochester ever since. She has a named professorship, the J. Lowell Endowed Chair in Biochemistry and Biophysics, and has concomitant appointments in pediatrics and in oncology. She is the founding director of the Center for RNA Biology. Her research has centered around understanding what happens in the cell during disease, especially concerning the role of RNA in sickness and health. She has demonstrated that a number of inherited as well as acquired conditions may be a consequence of pathological alterations in RNA and especially in mRNA. She discovered the so-called nonsense codon-mediated mRNA decay, the elimination of RNA transcripts containing nonsense codons. The nonsense codons are trinucleotide sequence units that do not code for any amino acid. (The "normal" codons code for amino acids, according to the genetic code.) A whole new kind of biomedicine is developing of late, and Maquat has been one of its pioneers.

However busy Maquat may have been with her multifaceted research program, she has always been dedicated to advocating for the provision of adequate resources to graduate women, resources she was denied in her own time. She was the founding chair and has remained the chair of Graduate Women in Science since 2003. This is an NIH initiative based on a predoctoral grant in cellular, biochemical, and molecular sciences; the program assists the professional and personal development of the graduate women at the Medical Center and the College of Arts and Sciences of the University of Rochester. It makes use of available expertise within the university and involve guest speakers not only from academia but also from industry and elsewhere. The areas addressed are broad and chosen with an eye to possible needs for broad-based instruction. Speakers represent large firms as well as start-ups, and different domains of activity, such as forensics, grant management, science writing, journal editing, and patent law.

A number of highly prestigious awards and distinctions have recognized Maquat's contributions to science and to assisting women graduates. The latest is her Wolf Prize in Medicine in 2021, shared with two others, Joan Steitz and Adrian Kraine, for their groundbreaking studies in RNA biology.

Katalin Karikó

Katalin Karikó (b. 1955) was born in Kisújszállás,[16] a town some 140 kilometers (85 miles) to the east of Budapest; her father was a butcher. She went to local schools and has praised her teachers there for inculcating in her the love of mathematics and the sciences. At the University of Szeged she majored in biochemistry and received the equivalent of a master's degree. She started her career with a scholarship at the Szeged Biological Research Center of the Hungarian Academy of Sciences and remembers the sizzling scientific atmosphere at the laboratory. The scholarship from the Academy

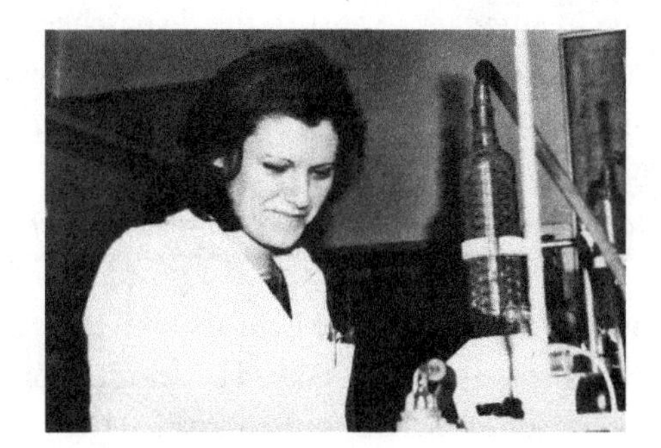

Katalin Karikó in 1980. Courtesy of Katalin Karikó.

[16] Karcag, a similarly small town, adjacent to Kisújszállás, is the birthplace of the Nobel laureate Avram Hershko.

Katalin Karikó in 2020. Courtesy of Katalin Karikó.

of Sciences allowed her to embark on her doctoral studies. Her mentor was Jenő Tomasz, an organic chemist whose project was sequencing an RNA molecule. He was pioneering RNA research in Hungary, and Karikó's dedication to this area of science traces back to his impact on her, a dedication that would prove invaluable during a future global crisis. Tomasz soon emigrated to the United States, where he worked at a variety of venues before retiring back in Hungary.

Karikó completed her doctorate in 1982 and stayed on at the Szeged Biological Research Center. There did not seem anything particularly remarkable about her performance there. The Center fell on hard times in the 1980s, and some of its associates were let go, among them Karikó. This was highly unusual under the socialist system, so the situation must have been very bad indeed. She received the news about her dismissal in 1985, on her 30th birthday. There was nothing she could do, and there was no other venue in Hungary to absorb the dispatched researchers. The state-run economy was undergoing a crisis, and there was no European Union then to bail out the country or provide employment for the surplus scientists.

The United States was thought to have ample opportunities, and following persistent applications, Karikó landed a postdoctoral position at the Department of Biochemistry of Temple University in Philadelphia. This position lasted for three years and was followed by another for two years at the Department of Pathology of the Uniformed Armed Services University of the Health Sciences in Bethesda, Maryland, near Washington, D.C., a federal medical school. Her project focused on interferons, which are signal proteins. When a cell is infected by viruses, it sends interferons to the neighboring cells to induce their antiviral defense. This project provided an opportunity for her to learn a great deal of the modern techniques of molecular biology.

In 1989, Karikó began a long-lasting, bumpy relationship with the University of Pennsylvania, a prestigious, private Ivy League school. She had a tenure-track position, which did not work out, but she preferred to continue on in a demoted position rather than leave. Not long before Karikó's COVID-19 vaccine fame, she was asked to give a talk at Harvard University, focusing on her failures. She remarked that this was easy, because she had plenty.

All Karikó's hard work was slow in demonstrating any concrete therapeutic promise. When she was in a so-called soft-money position, she needed outside support to continue; alas, such support was not forthcoming. At UPenn, she started at cardiology and then moved to neurosurgery. She always found enthusiastic colleagues who shared her vision, at least for a while.

Then came her long-awaited break, still at UPenn, in the form of a chance meeting with a recently appointed professor, Drew Weissman, at the Perelman School of Medicine. The encounter injected Karikó with new hope to continue her efforts. Weissman was a Brandeis MD/PhD who had done postdoctoral work with Anthony Fauci at the National Institute of Allergy and Infectious Diseases of NIH.[17] Weissman had ambitious plans to find remedies for several diseases. When Karikó told him her ideas about the use of mRNA, he recognized the unique potentials of her approach. A fruitful interaction developed between them.

Messenger RNA was discovered some 30 years before, and quite a number of scientists, among them some future Nobel laureates, acted as midwives in its discovery. None could be pointed to specifically as *the* discoverer or discoverers, and none received a Nobel Prize specifically for it. Considering its significance in hindsight, it would have been worthy of such a recognition. The molecule mRNA is a single-strand nucleic acid with the same sequence of nucleotides as the double-strand DNA from which it carries the genetic information to the ribosome. The function of mRNA explains why it is called a "messenger." The genetic information from the DNA is transcribed by means of an enzyme (a protein) called "RNA polymerase." According to the information the ribosome receives from the mRNA, it produces proteins. The ribosome is the protein factory; essentially it is the hardware of protein production, and the mRNA is the software. The

[17] Anthony Fauci is currently (2021) chief medical advisor to President Joe Biden.

entire process is called "translation," and it constitutes a part of the "central dogma," as labeled by Francis Crick, according to which, the information flow has a one-way direction:

DNA → RNA → Protein.

Never the reverse.[18] These basic principles, however, were far from suggesting any direct pharmaceutical applications.

In one of his immunology projects, Weissman was using DNA to develop a vaccine against HIV (the human immunodeficiency virus). When it did not work, Karikó suggested trying mRNA. The advantage of RNA over DNA is manifold. For DNA to bring about change, it has to penetrate the cell nucleus; for RNA, it suffices to enter the cytoplasm, that is, the cell material outside the nucleus. RNA acts more quickly in producing the necessary protecting proteins. It does not affect the genetic content of the target cells, whereas DNA may cause unwanted mutations. Once the DNA is brought into the cell nucleus, it stays, regardless of its efficacy. In contrast, RNA decomposes, so it is no longer there if it is not needed. In case its presence is needed because it has the desired efficacy, it can be augmented by providing additional amounts, just as with any treatment. Weissman followed up on Karikó's suggestion and injected mice with mRNA. The first results were discouraging; many of the mice suffered from inflammation and some died.

Weissman and Karikó together solved the problem with "clever biochemistry," as he called it.[19] Indeed, it originated with Karikó's deep knowledge of RNA chemistry. They replaced one of the four nucleotides with a similar building block whose participation did not alter the efficacy of the treatment but stopped the inflammation. In principle, they now had a reliable tool of using mRNA for creating vaccines for any disease. Aside from the emergency-approved COVID-19 vaccine, however, none has so far been approved. Still, this single success has so far saved millions and millions of lives. The work continues; currently there is an mRNA treatment being developed in Weissman's lab for sickle cell anemia. Karikó and Weissman jointly started a biotech company as a spinoff, but Karikó left it when she left UPenn and assumed a leading position at the BioNTech company in Mainz,

[18] This is no longer strictly the case. The enzyme called reverse transcriptase is capable of generating DNA from RNA.

[19] Drew Weissman, private email communication, 2021.

Germany, where her official title was executive vice president until in 2022 she left the company.

Karikó left Hungary under unpleasant circumstances. Had she stayed, as she put it in an interview, she would have become a mediocre, disgruntled researcher. However, she rose to world fame as the principal creator of a life-saving vaccine, lifting much of the weight of a global pandemic. She has become a great statesperson of mRNA science for medical applications. Her perseverance and stamina coming from her researcher's acumen brought hope to the world and appreciation to her life's work.

Paula T. Hammond

Paula Hammond, 2006. Photograph by Douglas A. Lockard, Science History Institute. Source: Wikipedia Commons.

Paula T. Hammond (b. 1963) is a chemical engineer who became famous for designing polymers (large molecules that consist of many repeating units) and nanoparticles (very small particles of about 1 to 100 nanometers) with the purpose of using them, for example, for drug delivery.

Hammond was born in Detroit, Michigan, to a PhD biochemist father and a nurse mother who had a master's degree. Hammond's chemistry teacher in high school inspired her love for chemistry. Thanks also to her good performance in math and science, she gravitated toward chemical engineering. She studied at MIT, where she was a rare phenomenon in chemical engineering: a Black woman among white men. She liked to face challenges rather than avoiding them, and this attitude has characterized her entire career.

Her academic performance and grace in social situations made her successful. She graduated from MIT with a bachelor's degree in 1984. She and her fellow engineer fiancé moved to Florida, where she had a job as a process engineer at a plant producing cell phones. Later, while working at the Georgia Institute of Technology, she continued her studies, and in 1988 she earned a master's degree. She returned to MIT for a PhD doing research on polymers, which she completed in 1993. Thanks to a National Science Foundation postdoctoral fellowship, she spent a year in George M. Whitesides's group in the Chemistry Department at Harvard University.

In 1995, she was appointed assistant professor of chemical engineering at MIT, where she built up her research group and developed its investigations into designing new polymers and nanoparticles with desired properties. The applications of her new materials range from medicine to energy production and fuel cells. She designed, for example, materials for drug delivery and for wound healing. She helped found institutes for the different possible applications of her technologies, among them, the Institute for Soldier Nanotechnologies, with the aim of improving protection for and survivability of the troops. One of the materials can be sprayed onto wounds to accelerate blood clotting; another application is a technology to build drug delivery films with alternating drug and polymer layers. Hammond also played a part in creating a genetically modified virus to function as a tiny battery, which is used for powering a variety of miniature devices.

Over the years, Hammond advanced both in science and in her academic positions. In 2015, already the David H. Koch Professor of Engineering, she was named the head of the Department of Chemical Engineering at MIT. As of 2021, her Wikipedia page lists some 50(!) honors and recognitions received between 1990 and 2019. No wonder she is one of the star professors of MIT.

Jennifer A. Doudna and Emmanuelle M. Charpentier

Emmanuelle M. Charpentier and Jennifer A. Doudna, 2017, on the cover of *Magyar Kémikusok Lapja*, December 2020. Courtesy of *Magyar Kémikusok Lapja*.

Two high-profile, internationally renowned scientists and their joint Nobel Prize presented a unique case: the first Science Prize for two women alone. The biochemists Jennifer A. Doudna (b. 1964) and Emmanuelle

M. Charpentier (b. 1968) have made many discoveries in biochemistry individually, but the Nobel Prize motivation singled out a joint discovery: the development of the method for genome editing.

Doudna was born in Washington, D.C., but grew up in Hawaii. Her father taught American literature at the University of Hawaii at Hilo and her mother taught history at the local community college. They had a home full of books, including popular science titles. These provided inspiration for Doudna, among them James D. Watson's *The Double Helix*. Additional inspiration came from her teachers at Hilo High School. She attended Pomona College in Clermont, California, majoring in biochemistry. There, again, she had inspiring science teachers. She earned a bachelor's degree in biochemistry in 1985 and went for her PhD at Harvard Medical School, which she earned in 1989 in the fields of biological chemistry and molecular pharmacology. The future Nobel laureate Jack W. Szostak was her mentor, and her dissertation conceived of a system to increase efficiency of a self-replicating catalytic RNA.

Charpentier was born in Juvisy-sur-Orge, France, a small community some 18 kilometers (11 miles) south of Paris. Its website now lists Charpetier among its top 10 notable people. For university studies she enrolled at the Pierre and Marie Curie University (today, the Faculty of Science of the Sorbonne), majoring in biochemistry, microbiology, and genetics, and completed her studies in 1992. For graduate school, she went to the Pasteur Institute (1992–1995), obtaining a doctorate investigating antibiotic resistance.

Following graduate school, Doudna worked in molecular biology at Massachusetts General Hospital and in genetics at Harvard Medical School. She then went as a postdoctoral researcher in biomedical science to the University of Colorado at Boulder (1991–1994), where the recent Nobel laureate Thomas Cech became her mentor. She moved to Yale University as an assistant professor in 1994 and was appointed Henry Ford II Professor of Molecular Biophysics and Biochemistry in 2000. One of her projects at Yale was the crystallization and determination of the three-dimensional structure of an RNA enzyme, a ribozyme, by X-ray crystallography. She spent the next year at Harvard's Chemistry Department as a visiting professor. Her latest move was joining the University of California, Berkeley in 2002 as a professor of biochemistry and molecular biology. Eventually she assumed the directorship of the joint Innovative Genomics Institute of UC Berkeley and UC San Francisco.

Charpentier has had an even more eventful career. After the Pasteur Institute, she crossed the Atlantic for a postdoctoral stint (1996-1997) at Rockefeller University in New York City. With Elaine Tuomanen, she investigated the utilization of mobile genetic elements for altering the genome of the pathogen *Streptococcus pneumoniae*. Between 1997 and 1999, Charpentier was a researcher at New York University Medical Center, and for the next two years at St. Jude Children's Research Hospital as well as the Skirball Institute of Biomolecular Medicine, both also in New York City. Her work concerned gene regulation and manipulation.

In 2002, she returned to Europe and held various positions at the University of Vienna, where she acquired habilitation—a sort of higher doctorate—a prerequisite for professorial appointment in much of Central Europe. During her concluding time in Vienna, 2006-2009, she was a laboratory head and associate professor at the Max F. Perutz Laboratories, a joint institution of the University of Vienna and the Medical University of Vienna. Over the course of the next decade, her affiliations and collaborations expanded significantly. Before her assignment in Vienna ended in 2009 she already had an appointment up and running at the Laboratory of Molecular Infection Medicine at Umeå University in northern Sweden, which she maintained until 2017. Meanwhile, she held professorial appointments in Germany in 2013-2015 at the Helmholtz Centre for Infection Research in Braunschweig and the Hannover Medical School. In 2015, she was invited to join the Max Planck Society and was appointed director at the Max Planck Institute for Infection Biology in Berlin. In 2018, she became the founding scientific and managing director of the Max Planck Unit for the Science of Pathogens in Berlin.

Both Doudna and Charpentier had become internationally renowned researchers with outstanding achievements before they met in 2011. They realized their overlapping interest from their publications. After meeting it was clear they had a personal rapport and decided to embark on a collaboration that immediately bore fruit. Here I attempt a simplified interpretation of their research. First, the term "CRISPR/Cas9," as it is often called, should be deciphered. CRISPR is the abbreviation of clustered regularly interspaced short palindromic repeats. These are DNA sequences in bacteria and single-cell animals lacking cell nuclei (archaea). Cas9 is a protein. Its deciphered name is CRISP-associated protein 9, a vital player in the immunological defense of certain bacteria, providing protection against viral DNA and plasmids. The combination CRISPR/Cas9 is a tool for editing genomes.

"Editing" here means the same as editing a manuscript: cutting, pasting, augmenting, or removing parts of the genome. CRISPR/Cas9 can be looked at as genetic scissors, a tool with which it is possible to rewrite the code of life.

It is possible to change the DNA of any living organism with high precision. By changing the DNA, inherited diseases can be cured, cancer therapies can be created. The potential applications are limitless. It may also have the potential of altering a living organism into something else, and this is where serious ethical considerations emerge. In the early 1970s, when genetic engineering appeared to be an imminent possibility, scientists were the first to declare a moratorium on further experiments before they unwittingly (or even worse, wittingly) created a monster. (See section on Maxine F. Singer.) Doudna seems particularly concerned about such eventualities and, along with other leading biologists, has called for a worldwide moratorium in the clinical application of CRISPR/Cas9 gene editing. She has supported the usage of this technique in somatic gene editing (that is, limited to a specific organism), but not in gene editing in which the alterations would be passed to the next generation—that is, not in germline gene editing.

Doudna and Charpentier were awarded the 2020 Nobel Prize in Chemistry for their contribution to the methodology of gene editing. They did not discover CRISPR/Cas9, but they developed it into a revolutionary tool. As is often the case, the Nobel Prize divides the haves from the have-nots. One only wonders why the possible third slot in this particular Nobel was left unfilled. One candidate might have been the Lithuanian biochemist Virginijus Šikšnys (b. 1956) of Vilnius University. It has been noted that the publication of his discovery appeared only months after that of Doudna and Charpentier. A few months may not be important in making a discovery, but it may be for priority issues. His manuscript was initially rejected by a publication, and the journal that finally printed it took several months to do so. In contrast, the manuscript by Doudna and Charpentier was reviewed right away upon submission and published within weeks.[20] A complicated patent dispute and court case had developed, and an MIT-Harvard group was awarded patent rights. It is not at all unprecedented that patent rights and Nobel Prizes involving the same discoveries are awarded to different people. A famous

[20] In 2018, Emmanuelle Charpentier, Jennifer A. Doudna, and Virginijus Šikšnys received the Kavli Prize in Nanoscience "for the invention of CRISPR-Cas9, a nano-tool used for editing DNA. CRISPR-Cas9 has numerous applications to biology, agriculture, and medicine through the modification of genes." (Winners of 2018 Kavli Prizes Announced | Philanthropy news | PND (philanthropynewsdigest.org))

case is the patent rights and the Physics Nobel Prize in 1964 for the discovery of the laser.

There is, however, no dispute about the enormous significance of the CRISPR/Cas9 methodology. It is also true that its application may lead to very different outcomes. This is not the first time that science has provided a tool that can be used to the benefit of humankind or its destruction. Choices of use and application are not in the hands of scientists but in the hands of society. Scientists can help society in reaching the right choices by helping to cultivate and share in an *informed* society.

Carolyn R. Bertozzi

Carolyn R. Bertozzi lecturing in 2011 at the Technical University of Darmstadt. Photograph by Armin Kübelbeck, public domain, Source: Wikimedia, https://commons.wikimedia.org/wiki/File:Carolyn_Bertozzi_IMG_9384.jpg.

Carolyn R. Bertozzi (b. 1966) of Stanford University shared the 2022 Nobel Prize in Chemistry with Morten Meldal of the University of Copenhagen and K. Barry Sharpless of Scripps Institute, La Jolla, "for the development of click chemistry and bioorthogonal chemistry." Meldal and Sharpless worked out a technique for snapping molecular building blocks together in an extremely efficient way. It is called "click chemistry." The word "click" is best known for referring to pressing a button or touching the screen of the computer. It signifies a quick action. Click chemistry refers to many chemical reactions in which small units are joined to generate various substances mimicking processes in nature. Bertozzi built on this technique in developing new chemical reactions induced inside of the living system. For her new chemistry, she coined the name "bioorthogonal chemistry" in 2003. With time, its applications in biomedicine and therapy are ever broadening.

Bertozzi and her two sisters grew up in Lexington, Massachusetts. Their MIT physics professor father and opportunities at MIT camps helped them develop their academic interests. Bertozzi was gifted not only in science but also in music. She received her A.B. degree at Harvard University and her PhD in chemistry at University of California, Berkeley in 1993. Following a postdoctoral fellowship at University of California, San Francisco (1993–1996) and her Faculty positions at Berkeley (1996–2015), she has been at Stanford University. She is an innovative chemist, always with an eye on solving problems for biomedicine and therapy. One of the areas of her activities has been glycobiology, the biology of saccharides (sugars). They constitute a large class of biologically important substances that used to be considered "less glamorous" than nucleic acids and proteins. Bertozzi's achievements have helped the saccharides to be elevated into another focal point of academic and medicinal interest. Her Wikipedia entry as of October 5, 2022, the day her Nobel Prize was announced, listed 50 awards and honors, some of them among the highest.

6
Biologists and Biomedical Scientists

Traditionally, the areas of science where women have had the best opportunities were crystallography, chemistry and biochemistry, and biology and the medical sciences, though even in these areas the road to success has often proved bumpy.

Zinaida V. Ermoleva

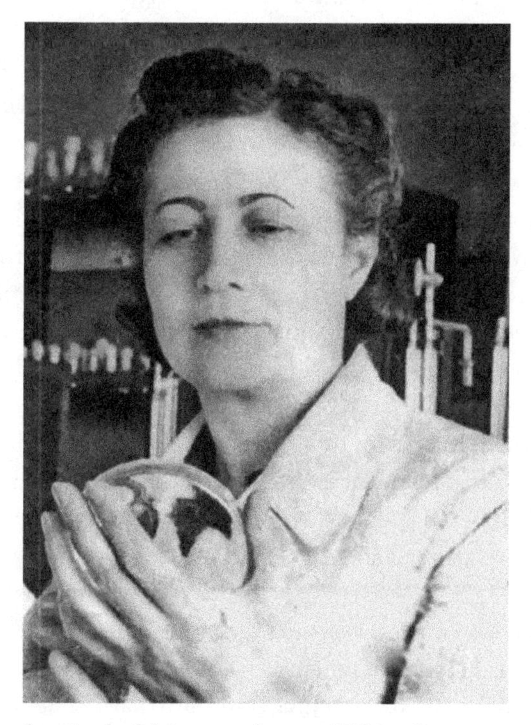

Zinaida V. Ermoleva in the laboratory. Source: Wikipedia.

Zinaida V. Ermoleva (1898-1974) was born in Frolovo, Volgogradsky Region, Russia. Her father descended from the Don Cossacks. There were seven children in the family; five boys and two girls. The father died in 1909 and the mother brought up the children alone. Zinaida attended the girls' high school in Novocherkassk. According to family legend, she read a book about the great Russian composer Pyotr Tchaikovsky and learned that he died of cholera, at which point she decided to devote her life to finding a cure for this terrible illness. She became interested in microbiology during her sophomore year in the medical school in Rostov-on-Don and started doing voluntary work in a microbiology laboratory. After graduation in 1921, she was appointed assistant in the Department of Microbiology.

In 1922, an epidemic of cholera broke out in the region of the River Don and Ermoleva was charged with heading up a bacteriological division. She isolated a pathogen supposed to be the carrier of cholera. In order to be sure, she mixed it in water and drank it. She became severely ill with cholera, but in possession of this evidence, she introduced strict regulations about sterilizing water with chlorine and saved many lives.

Three years later, she was assigned to Moscow at an institute that in a few years' time became part of the All-Union Institute of Experimental Medicine. She continued her studies in microbiology and was considered to be one of the preeminent experts in epidemiology in the Soviet Union. In 1942, she was sent to Stalingrad, where her efforts saved the Soviet troops and the locals from a cholera epidemic, which had already spread among the German troops. During the war, she was charged with producing penicillin. In 1944, a group of Western scientists visited her laboratory. Howard Florey, a pioneer of penicillin, compared the British and Russian penicillin samples, expressing his admiration for Ermoleva and calling her "Lady Penicillin."

Following the war, she continued her studies of antibiotics and headed laboratories and institutes concerned with antimicrobial materials of natural origin. During the last two decades of her life she was in charge of the Microbiology Department at the Central Post-Graduate Medical Institute in Moscow.

She was married for a few years to the renowned virologist Lev A. Zilber.[1] Although they had divorced, she stood by him at the time of his unjust incarceration. Selflessly, she did everything possible to have Zilber freed, which he was in 1944. Her second husband, the microbiologist Aleksei A. Zakharov, was also arrested on false charges, and all her attempts to have him freed

[1] Istvan Hargittai and Magdolna Hargittai, *Science in Moscow: Memorials of a Research Empire* (Singapore: World Scientific, 2019), 198–199.

failed. Although she was informed that he had died in prison in 1940, he was actually executed in 1938.

The tragedies in her life became known only following the collapse of the Soviet Union, and it has since inspired films, two TV series, and a novel. She has become a role model for young women in Russia who choose microbiology for their profession.

Barbara McClintock

Barbara McClintock, 1947, at her bench at the Cold Spring Harbor Laboratory. Source: Wikimedia Commons.

U.S. postage stamp honoring Barbara McClintock, 2005.

Barbara (born Eleanor) McClintock (1902–1992) was a dedicated and determined scientist who painstakingly used the classical tools of genetics available to her at the time and made seminal discoveries about gene mobility and its manifestation. Her discoveries were ahead of contemporary science. When progress in science finally caught up with her contributions, she received her long-deserved Nobel recognition.

She was born in Hartford, Connecticut, as the third child of four. Her father, Thomas H. McClintock, was a homeopathic physician, and her mother was a homemaker and a descendant of one of the *Mayflower* families. They changed her name from Eleanor to Barbara when they saw her independent character developing and thought that Eleanor was too feminine for her. In later life, McClintock identified her independence with the ability to survive alone, as she did all her life. When the family moved to Brooklyn, McClintock attended Erasmus Hall High School on Flatbush Avenue, from which she graduated in 1919.[2] She continued her studies at the College of Agriculture (today, College of Agriculture and Life Sciences) at Cornell University. Her mother opposed this move, fearing that a college education would make McClintock unmarriageable, but her father supported her plans.

McClintock received her degrees in botany: a bachelor's in 1923, a master's in 1925, and the PhD in 1927. However, when she took a course in genetics by Claude B. Hutchinson (1885–1980), it changed her outlook. He sensed her intense interest and invited her to a graduate course in genetics while she was still an undergraduate. From the start of her research career McClintock focused on cytogenetics, studying how the structure and morphology as well as the function of chromosomes influence inheritance. She worked out the necessary methodology for her research, for example, the staining of the chromosomes to make them visible. She chose maize for the plant of her inquiry. She succeeded in demonstrating that specific chromosomes were responsible for inheriting certain traits.

From the outset, her work gained recognition and landed her postdoctoral fellowships from the National Research Council. She continued at Cornell for some time before moving to the University of Missouri, where she learned the technique of introducing mutations with X-rays. With mutagenized maize she made novel observations on their chromosomes. She had a powerful research tool, which accelerated the occurrence of mutagens over the level that natural mutagenesis would provide. Her discoveries were rather forward pointing; suffice it to mention that she was thinking about the tips of chromosomes being protected by telomeres whose composition was not yet known. Such telomere protection figured prominently in discoveries

[2] Her high school graduated many notable alumni, among them another future Science Nobel laureate, Eric Kandel. Erasmus Hall High School closed in 1994 due to poor performance and was broken up into a set of smaller schools.

recognized by the 2009 Nobel Prize (see below on Elizabeth H. Blackburn and Carol W. Greider). McClintock was supported by the Guggenheim Foundation and the Rockefeller Foundation. She wanted to augment her postdoctoral training in Germany and planned to spend the academic year 1933-1934 there. However, her would-be mentor had to flee the country in the wake of the Nazi takeover, and McClintock's visit was cut short.

She hoped that the University of Missouri might offer her tenure, but her hopes faded when she was not included in faculty meetings and was left uninformed about events that she should have attended. Following another visiting position at Columbia University, and other temporary assignments, she finally settled at a full-time research position at the Cold Spring Harbor Laboratory of the Department of Genetics at the Carnegie Institution of Washington, as it was then called. This turned out to be an ideal venue for her long-term projects where she could work without outside pressure, peer interference, and tensions of competition.

In 1944 she was elected a member of the National Academy of Sciences, which many consider the most important recognition because it is given by one's peers. The significance of her election was enormously enhanced by her being a woman—she was only the third female scientist accorded this honor. The first was Florence R. Sabin in 1925 (see chapter 7) and the second, Margaret F. Washburn, in psychology and cognitive science in 1931. The gaps in years, 1925-1931-1944, only add weight to the significance of these elections.

At the Cold Spring Harbor Laboratory McClintock continued her broad-based research program on the maize chromosomes. Lacking an X-ray apparatus, she designed other tools to induce mutagenesis. She conducted systematic studies on the mechanism of the color patterns of maize seeds and their inheritance.

In 1948, she made surprising discoveries about the mobility and certain instances of interchangeability of genes. Between 1948 and 1950, she hypothesized about the mobile elements in the genetic setup. She investigated how genetic regulation made it possible for complex multicellular organisms to produce cells of identical genomes but of different functions. She published on and lectured about the mobile genes, and tried to get through to her colleagues with her admittedly revolutionary ideas and observations. This was not easy, and at some point she felt that her efforts were becoming counterproductive. In 1953, she ceased publishing for some time.

It was a welcome change, therefore, when the National Academy of Sciences provided her funding for an extensive study of the evolution of maize in Central America and South America. In 1967, she retired officially from her position but continued training graduate students and working with associates as a Distinguished Service Member of the Carnegie Institution, using her laboratory at Cold Spring Harbor as a scientist emerita.

From the 1960s onward, the world of science started catching up with her discoveries. Justification and approval was coming out of the work of such greats as François Jacob and Jacques Monod of the Pasteur Institute in Paris. McClintock herself noticed similarities between recent discoveries and her findings decades before and wrote about them. Accolades started pouring in. One of the first was the U.S. National Medal of Science, given to her by President Richard Nixon in 1970. Further prestigious recognitions were the Albert Lasker Award for Basic Medical Research, the Wolf Prize in Medicine, and a good number of others, before the pinnacle of recognition, the Nobel Prize in Physiology or Medicine in 1983 "for her discovery of mobile genetic elements." She was the first woman to receive an unshared prize in this category and the first American woman receiving an unshared prize in any category. She was compared to Gregor Mendel. She was in good company when in 2005 the U.S. Postal Service issued a set of four 37-cent stamps with the portraits of four scientists: McClintock, John von Neumann, Josiah Willard Gibbs, and Richard Feynman.

Aleksandra A. Prokofieva-Belgovskaya

Aleksandra A. Prokofieva-Belgovskaya (née Prokofieva, 1903–1984) was born in Aleksandrov, Vladimir Region, on the so-called Golden Ring, a major tourist destination in Russia. She graduated from a Petrograd pedagogical college in 1923 and taught in Leningrad schools until 1930.[3] Afterward she worked in biological research institutes of the Soviet Academy of Sciences, and from 1933 in the Institute of Genetics, headed at that time by Nikolai I. Vavilov.[4] There she worked with international scientists in Hermann

[3] Petrograd and Leningrad were names of the same city, which used to be called, and is today again called, St. Petersburg.

[4] Nikolai I. Vavilov (1887–1943) was an internationally renowned biologist-geneticist who perished in Stalin's prisons in 1943 on false charges and was exonerated decades later.

Aleksandra A. Prokofieva-Belgovskaya. Source: Wikipedia.

J. Muller's laboratory devoted to mutagenesis and learned the modern techniques of research from Calvin Bridges.[5] Prokofieva-Belgovskaya continued her research and presented her results in a dissertation for the higher doctorate, DSc, in 1948. Her principal interest was cytogenetics—the science of the relationship between chromosomes and inheritance. The year was most unfortunate for her science; it was the culmination of the reign of the charlatan dictator of Soviet biology and agriculture, Trofim D. Lysenko. He conducted his ruthless campaign against the science of genetics and inheritance with the full support of Soviet dictator Joseph Stalin. Prokofieva-Belgovskaya could not receive her degree, was fired from the Institute of Genetics, and for years could not continue her research in genetics.

[5] Hermann J. Muller (1890–1967) was an American geneticist who worked for years in the Soviet Union but returned to the United States sensing the growing danger for his science and scientists. He received the Nobel Prize in Physiology or Medicine for the discovery of mutagenesis by X-ray irradiation. Calvin Bridges (1889–1938) was an American geneticist who spent four months visiting Nikolai Vavilov's Institute in 1935.

She found employment in a ministerial research institute and made valuable contributions to the industrial production of antibiotics. There was an improvement in her situation in 1956 when she obtained a position in an institute of the Academy of Sciences where she could resume her research in genetics. Officially, such research was still banned in the Soviet Union, but it could be conducted under the pretext that she studied the variations in species under radioactive irradiation within the framework of the nuclear program. In 1962 she joined Vladimir Engelhardt's laboratory in what would later become the Institute of Molecular Biology. While continuing her work with Engelhardt, in 1964 she was offered the opportunity of initiating a new laboratory of cytogenetics at a research institute of the Academy of Medical Sciences. In time, this laboratory merged with the Institute of Medical Genetics and she carried on heading her laboratory to the end of her life. She did this without remuneration after she retired. She also kept her position at the Institute of Molecular Biology—later, the Engelhardt Institute of Molecular Biology—until the end of her life.

Rita Levi-Montalcini

Rita Levi-Montalcini (1909–2012), the corecipient of the 1986 Nobel Prize in Physiology or Medicine, did not consider herself a scientist: "I don't believe that I have ever been a scientist. I believe that my approach to science

Rita Levi-Montalcini with Viktor Hamburger, 2000. Courtesy of Rita Levi-Montalcini.

Rita Levi-Montalcini in her home in Rome. Photograph by M. Hargittai.

was from the point of view of the beauty of the nervous system and not just plainly because I was interested. Still now, I don't believe I am a scientist; I approach science more from an artistic point of view than from a scientific one."[6]

Learning about her family background helps us to understand her surprising statement. She was born in Turin, Italy, into a well-to-do Jewish family, in which her twin sister, Paola, and their mother were talented painters and her brother was an architect. Her father, as head of the family, made all the decisions, and he strongly opposed professional careers for women. He decided that since it was too difficult to find the balance between family, children, spouse, and a profession, women should not get a higher education. The young Rita was furious and decided that she would never marry; she only wanted to study. When she turned 21 years old, she enrolled at the medical school in Turin. Her father died just one year after she started her university studies. Their relationship is best described in the dedication to her autobiography, *In Praise of Imperfection*: "To Paola and the memory of our father whom she adored while he lived and whom I loved and worshiped after his death."[7]

[6] Magdolna Hargittai, "Rita Levi-Montalcini," in *Candid Science*, vol. 2: *Conversations with Famous Biomedical Scientists*, ed. Istvan Hargittai and Magdolna Hargittai (London; Imperial College Press, 2003), 367.

[7] Rita Levi-Montalcini, *In Praise of Imperfection* (New York: Basic Books, 1989), translation by Luigi Attardi from the original Italian *Elogio dell'imperfezione*.

Levi-Montalcini was one of seven girls among 300 students at the medical school. Two of her colleagues were Salvador Luria and Renato Dulbecco, both future Nobel laureates. All three of them were the students of Professor Giuseppe Levi (no relation to her), a well-known anatomist and histology professor at the University of Turin. The research direction of his laboratory, the development of the nervous system, fascinated Levi-Montalcini, and it became her lifelong interest. There she learned the very useful silver-staining technique, which made nerves visible under the microscope.

She graduated in 1936 and started a three-year course in neurology and psychiatry, but then history interfered. By 1938, Benito Mussolini's Fascist Italy had drawn close to Hitler's Nazi Germany, and the new Italian anti-Semitic laws excluded Jews from the university. In early 1939, Levi-Montalcini went for a few months to Brussels to the neurological laboratory, but in December, just before the Germans invaded Belgium, she returned to Italy. She set up a small laboratory in her bedroom and started experiments in neuro-embryology with chick embryos. In this, she was influenced by an article she had read a few years before by Viktor Hamburger (1900–2001), who worked at Washington University in St. Louis, Missouri. In 1941, due to the heavy bombing of Turin, the Levi-Montalcini family moved to the country, to Piedmont. There, she continued her experiments, for which only rudimentary means were needed. She bought fertilized chicken eggs from the farmers for her experiments.

In the autumn of 1943, after the German invasion of Italy, the family moved to Florence and went into hiding. Beginning in late 1944, after the region had been liberated, she worked as a medical doctor at the Anglo-American Headquarters. Finally, in May 1945, the family returned home and Rita resumed her position at the university as an assistant to Professor Levi, who became chair of the Department of Anatomy. At the end of the war, she and Professor Levi published their joint studies in a Belgian periodical. Hamburger read the paper, which disagreed with some of his hypotheses, and invited her to St. Louis to work on neuro-embryology. She arrived in 1947, intending to stay for a few months; she stayed for more than two decades.

First, she proved that her findings were correct. Soon after, Hamburger showed her an article about a large outgrowth of fibers caused by a malignant mouse tumor implanted into chick embryos. She repeated this experiment using the silver-staining technique and found that nerve fibers appeared everywhere in the embryo's organs. She hypothesized that the tumor released some agent which induced this fiber growth.

She presented her findings at the New York Academy of Sciences in 1951, but they did not stir any serious interest. Continuing her work, she used a new tissue culture technique by Hertha Meyer, a German Jewish refugee at the Institute of Biophysics in Rio de Janeiro. Levi-Montalcini visited Meyer, and using Meyer's technique, proved that Levi-Montalcini's hypothesis was correct: the tumors released a growth factor in the culture medium. This is what later became known as the nerve growth factor (NGF). At that time, another scientist, Stanley Cohen, joined the Hamburger laboratory, and an exceptionally fruitful collaboration followed. Rita Levi-Montalcini and Stanley Cohen shared the 1986 Nobel Prize in Physiology or Medicine for discovering the growth factors.

Levi-Montalcini returned to Italy during the 1960s. She was appointed director of the recently founded Institute of Cell Biology of the Italian National Research Council in Rome. Even after her official retirement, she kept going to the Institute and participating in its research program. At the time of my visit there, in 2000, the main research topic was still NGF. She said that its identification about half a century earlier was just the beginning. Her associates were investigating the peripheral and central nervous system and the immune and endocrine systems. The new results showed promise that this molecule might be used to treat neurological and psychiatric disorders, for example, Alzheimer's disease, dementia, schizophrenia, depression, and autism. It has also been shown to speed up wound healing, and it might be used to treat skin ulcers. Scientists at the University of Pavia found that young people who recently fell in love show a much higher level of NGF than those who are not in love or those who have been for a long time. The NGF level stays high for about one year.

Rita Levi-Montalcini passed away in 2012, at the age of 103. When she was 100, some reports in the media supposed that she must have been taking NGF for years in the form of eye drops. Whether this was the case and it increased her longevity, we can't know. NGF may help neurons in the brain to survive. At the time of her centennial, she stated that her mental capacity was better than decades before. She was beautiful and radiated elegance, and this seemed not to change with age. She was petite, thin, but strong and purposeful. She was a Senator for Life in Italy and she took her assignments seriously. In 2006, when she was 97, she voted against decreasing funding for science and thus saved support for science.

Her poetic nature comes to life in her book when she talks about NGF and the 1986 Nobel ceremonies in Stockholm:

> It was in the anticipatory, pre-Carnival atmosphere of Rio de Janeiro that in 1952 NGF lifted its mask to reveal its miraculous ability to cause the growth, in the space of a few hours, of dense auras of nervous fibers. Thus began its saga. . . . On Christmas Eve 1986, NGF appeared in public under large floodlights, amid the splendor of a vast hall adorned for celebration, in the presence of royals of Sweden, of princes, of ladies in rich and gala dresses, and gentlemen in tuxedos. Wrapped in a black mantle, he bowed before the king and, for a moment, lowered the veil covering his face. We recognized each other in a matter of seconds when I saw him looking for me among the applauding crowd. He then replaced his veil and disappeared as suddenly as he had appeared. . . . Will we see each other again? Or was that instant the fulfillment of my desire of many years to meet him, and I have henceforth lost trace of him forever?[8]

Frances O. Kelsey

" 'Heroine' of FDA [Food and Drug Administration] Keeps Bad Drug Off the Market"[9] is the headline to a front-page story in the Sunday, July 15, 1962, issue of the *Washington Post*. Similar stories appeared in other major newspapers in the United States. Horror stories about thalidomide—in Europe it was called Contergan—filled newspapers all over the world in the early 1960s. The drug was developed in the 1950s by a company in Germany as a sedative and for relieving morning sickness of women in the first trimester of their pregnancy. Children whose mothers took this drug during pregnancy were born with severe birth defects, mostly deformed or missing limbs. About 10,000 children were affected in Europe, Canada, and other parts of the world. In contrast, only very few cases occurred in the United States, thanks to a conscientious and dedicated scientist at the Food and Drug Administration: Frances Oldham Kelsey (1914–2015).

[8] Magdolna Hargittai, "Rita Levi-Montalcini," 374–375.
[9] Morton Mintz, "'Heroine' of FDA Keeps Bad Drug Off the Market," *Washington Post*, July 15, 1962.

Frances O. Kelsey with President John F. Kennedy, 1962, at the White House when she received the Distinguished Federal Civilian Service Award. Courtesy of the John F. Kennedy Presidential Library, Boston.

Frances O. Kelsey at the time of my visit with her, 2000. Photograph by M. Hargittai.

Frances Oldham was born in Cobble Hill, a village on Vancouver Island in British Columbia, Canada. She finished high school at the age of 15 and went to McGill University in Montreal, where she received her BSc. She had a choice of continuing her studies or joining the "breadline," because finding a job was next to impossible—this was the time of the Great Depression. She stayed at McGill, in pharmacology, and received her master's degree in 1935. Her studies continued at the University of Chicago, at its newly founded Pharmacology Department. One of her projects was related to a request from the FDA, concerning a new drug called "Elixir Sulfanilamide." Sulfanilamide as a pill had been used for treating bacterial infections for quite some time with excellent results; it was considered a wonder drug. Now the manufacturer wanted to prepare it in liquid form as well, so that children would take it more easily. The company dissolved the drug in a chemical and, without further testing it, put it on the market, but this new form of the drug killed a number of patients. The FDA asked Oldham's professor to find out what caused these tragedies. Oldham found that the solvent, the toxic diethylene glycol, was the culprit. This study led to the Federal Food, Drug, and Cosmetic Act in 1938. This new law required that before a drug is placed on the market, its manufacturers have to prove its safety based on animal experiments, chemical experiments, and clinical studies. Oldham received her PhD in the same year.

One of her colleagues in the Pharmacology Department at Chicago was F. Ellis Kelsey, whom Frances would marry in 1943. After finishing her PhD, she went to the University of Chicago Medical School and got her medical degree in 1950. Their two daughters were born during this period. The family moved from place to place, following Ellis's job opportunities, and in 1960 they moved to Maryland. In August 1960, Frances started working at the FDA as its new medical reviewing officer. Investigating the thalidomide application landed on her desk.

By this time, Contergan had been used widely in European countries, mostly Germany, and because of its European acceptance, it might have sailed through the FDA without any hiccups. However, Kelsey and her chemist and pharmacologist colleagues noticed numerous problems as they examined the documentation submitted under the brand name Kevadon by the Richardson-Merrell pharmaceutical company. The animal tests were not properly recorded, and neither were the clinical studies. Furthermore, it was not indicated which chiral form of the molecule was tested, or perhaps it was the mixture of the two chiral versions.

Chirality is a special characteristic of the molecules of many substances. It is "handedness," meaning that there may be two versions of molecules of the same substance. The two versions are mirror images of each other, but they are not superposable, just like our left and right hands. The two versions are called "enantiomers." It is possible that while one of the enantiomers is a cure for a disease, the other may be a poison.

Thalidomide was known to exist in two enantiomers. It had been supposed that while one of the two provided the effects for which it could be used as a medication, the other form was harmless. But it turned out that one of the two enantiomers was teratogenic, causing birth defects. Unfortunately, the teratogenic effect was not limited to one of the two versions of the molecule, because the two versions rapidly interconvert into each other in the organism. But this was learned only much later.

As Kelsey continued her investigation, she requested more tests and additional documentation from the company. It might have appeared as if Kelsey was purposely extending the time of her study, and she was feeling increasing pressure to complete it. Then the news started coming from Europe about the birth defects and their possible connection to Contergan consumption. At this point Kelsey was satisfied not to allow thalidomide into the U.S. market. In the end, the thalidomide tragedies had significant impact on U.S. drug safety legislation. Kelsey's work helped to produce a regulatory regime—the Kefauver-Harris Amendments of 1962 and the Investigational New Drug Regulations of 1963—with stronger regulatory properties than any of the bills previously under discussion in Congress.

As head of the FDA's Division of New Drugs and later as director of the Division of Scientific Investigations, Kelsey was put in charge of ascertaining whether or not the suggested changes had been implemented. She retired from the FDA in 2005; by then, she was 90 years old. She passed away at the age of 101.

Back in 1962, due to the extensive media coverage, Frances Kelsey suddenly became famous in the United States. A national Gallup poll put her among the 10 most admired women in the world, in the same group of such celebrities as Jacqueline Kennedy and Queen Elizabeth II. She received the President's Award for Distinguished Federal Civilian Service from John F. Kennedy, and in 2000 she was inducted into the National Women's Hall of Fame. Exactly 50 years after she started to work on the thalidomide case, the FDA created the Dr. Frances O. Kelsey Award for Excellence and Courage in

Protecting Public Health, to be given annually to an FDA employee. The first recipient of the award was 96-year-old Frances O. Kelsey herself.

Anne McLaren

Anne McLaren in her laboratory in Cambridge, UK, 2004. Photograph by M. Hargittai.

Sometime in 2003, Anne McLaren (1927–2007) was invited to a meeting on stem cell research convened by the Pontifical Academy in the Vatican. While all the other invitees talked about their science and about the potential medical applications of stem cells, she talked about the ethical and political implications of the research. She had such a great impact that the laws for in vitro fertilization have since become stricter in Italy. The noted stem cell biologist Helen M. Blau of Stanford University remembered her in her obituary: "The image of this tiny intrepid woman, who spoke her mind despite facing an impossible task, will remain with me always."[10]

[10] H. M. Blau, "Anne McLaren (1927–2007)," *Differentiation* 75 (2007): 900.

McLaren was the most conscientious ambassador of science whenever there was a need for this in public debates or official hearings on the question of human reproductive technologies. At the same time, she was a very successful geneticist, whose research led to the possibility of in vitro fertilization. She was the fourth child of Christabel MacNaughten and Henry McLaren, 2nd Baron Aberconway, a Liberal Party member of Parliament and successful businessman. Until World War II, they lived in London, near Hyde Park. At the start of the war the family moved to their large estate in northern Wales. Anne finished her basic education in a private school in Cambridge by the time the war ended. She was hesitant about the direction of her further education and chose biology without giving much importance to this selection. For her first two years at Oxford, she studied (or, as the English say, read) zoology, physics, and math, and in the end she decided that zoology interested her most.

During her studies at Oxford she had famous mentors, such as J. B. S. Haldane and Peter Medawar. She was especially impressed by the work of the geneticist E. B. Ford and she decided to work in genetics. Another student there, Donald Michie, was also fascinated by this field and they started doing research together. She received her PhD in zoology in 1952. She and Michie received a grant to work at University College London. They moved there, and married the same year.

Their joint work turned out to be very successful. They investigated maternal effects between two inbred strains of mice, using embryo transfers between the two strains. They found that the embryos that were transferred resembled their uterine foster mothers and not their genetic egg mothers. They used a mouse colony for their experiment that grew so large over the years that they had to move it to the Royal Veterinary College in 1955.

McLaren did her most famous work in collaboration with the reproductive biologist John D. Biggers, which sounded deceptively simple in her description: "John Biggers was there in the next room and we got to know him very well. So the time came when he cultured the embryos and I transferred them into the uterus and we got mice born. That was the first time that embryos kept outside the body for 24 hours had been successfully reared into adulthood."[11]

[11] M. Hargittai, "Anne McLaren: Developmental Biologist," in *Women Scientists: Reflections, Challenges, and Breaking Boundaries* (New York: Oxford University Press, 2015), 141–142.

The work made headlines, but the social and ethical implications of this discovery started to sink in only gradually. In 1978, as a derivative of this research, the first "test-tube baby" was born. McLaren was invited to participate in committees and debates where various implications of in vitro fertilization were discussed. The most important of these was the so-called "Warnock Committee," which was appointed by the British government. It was charged with studying the social, ethical, and legal implications of human-assisted reproduction. The chair of this committee, Mary Warnock, was a well-known philosopher and writer. McLaren was the only member with scientific expertise in the topic. She remembered, "The Warnock Committee was set up in 1982 and it was a mixed committee with theologians and doctors and lawyers, different sorts of people. I was the only biologist in that particular area. Mary Warnock was a very good chair for the committee and we produced recommendations, advice to the government on possible legislation, which included recommendation that the government should set up an authority to regulate IVF (*in vitro* fertilization), both clinically and also human embryo research."[12]

McLaren's role was essential in formulating the guidelines that eventually led to the Human Fertilization and Embryology Act. Soon the Human Fertilization and Embryology Authority was founded, on which she served for 10 years as an active member. In later years she was equally active in debates about stem cells. She was also a cofounder of the Frozen Ark Project to collect and store the DNA of animals that are close to extinction. Despite her increased involvement in advising on scientific issues, the intensity of her research never diminished. In 1959, she moved to Edinburgh to the Institute of Animal Genetics, where she remained for 15 years. In 1974, she was appointed director of a new Mammalian Development Unit of the Medical Research Council at University College London and moved back to London. She retired from that position in 1992 and moved to Cambridge to the Wellcome Trust/Cancer Research UK Gurdon Institute. There, she continued her research until the day she died. Her latest interest was in mammalian primordial germ cells.

She summarized her work for me thus: "I have always enjoyed my work and I have worked on a number of different things because one thing has always led to another. Now, I am working on the imprinting of genes,

[12] M. Hargittai, "Anne McLaren: Developmental Biologist," in *Women Scientists: Reflections, Challenges, and Breaking Boundaries* (New York: Oxford University Press, 2015), 140–144.

which happens in germ cells. My primordial germ cell work led on to this imprinting of genes and also to stem cells which can be made from primordial germ cells. But I wouldn't single out any particular bit of my work as more important."[13]

In spite of her aristocratic roots, she was a socialist all her life, and so was Michie. They joined the Communist Party during the Cold War and supported Soviet and Eastern European scientists. She was a Fellow of the Royal Society (1975) and became its foreign secretary in 1991—the first time a woman had held this position. In 1993, she was president of the British Association for the Advancement of Science. She was made Dame of the British Empire in 1993, and in 2002 she received the Japan Prize, sharing it with Andrzej K. Tarkowski of Poland, for their pioneering work on mammalian embryonic development.

Her talent in science communication was not limited to the debates in committees and hearings. She excelled in sharing science with the general public in a way that was powerful and accessible. She said, for example, that when the embryo is outside the woman's body, the father and mother have equal rights, according to genetics. However, when it is inside the body, the woman's right is paramount, according to physiology.

McLaren was involved with women's rights and all women's causes. Although she felt herself fortunate in that she never experienced discrimination, she was very much aware that it existed. She was a charter member and the president of the Association for Women in Science and Engineering in Britain. She made it a point that in her laboratory there should be an equal number of men and women associates.

She had three children, each born two years apart. It never occurred to her to stop working. Her children became professionals. When I asked her what was most difficult about being a mother *and* a scientist, she answered, "Time. Time. Organization of time." Although she and Michie divorced in the late 1950s, they remained good friends. After their successful work on environmental effects on embryonic development, Michie left genetics. During World War II, he worked on cryptography at Bletchley Park and was one of the leaders of the project. He was a good friend of Alan Turing, the famous cryptographer, computer scientist, and creator of the "Turing machine." After the war Michie had become interested in genetics and that was

[13] M. Hargittai, "Anne McLaren: Developmental Biologist," in *Women Scientists: Reflections, Challenges, and Breaking Boundaries* (New York: Oxford University Press, 2015), 142–143.

how he met Anne. But code breaking and artificial intelligence were his genuine interests.

In early July 2007, McLaren and Michie were driving en route to Edinburgh, where Michie was to receive an award. In a tragic automobile accident, they both died.

Christiane Nusslein-Volhard

Christiane Nusslein-Volhard at the University of Tübingen, 2001. Photograph by M. Hargittai.

Christiane Nusslein-Volhard (née Volhard, b. 1942) was born in Magdeburg, Germany. Her parents did not have an academic background, but they supported all four of their daughters and one son in pursuing their academic interests. In Christiane's case, this meant science. She has pleasant memories from her high school years. The biology teacher was especially good in presenting interesting subjects, such as genetics, evolution, and animal behavior. This inspired Christiane to pursue higher studies in biology, first at Frankfurt University, then at Tübingen University, where they had just started a new biochemistry course. She also attended courses in genetics and microbiology.

Christiane graduated with a master's degree equivalent in biochemistry in 1968 and continued in the same city at the Max Planck Institute, where she received her PhD in 1974. Developmental biology seemed to be an exciting future path, an area of biology investigating how particular structures are formed, based on a pattern, for example, how cells grow and differentiate. She chose *Drosophila* (the common fruit fly) as the subject for her studies and teamed up with the American Eric Wieschaus at the European Molecular Biology Laboratory in Heidelberg. They investigated *Drosophila* mutants, a favorite subject for developmental biologists because the *Drosophila* larvae have 14 seemingly equal segments, yet they develop into different parts of the mature fly. Volhard and Wieschaus determined which genes were responsible for which mutations and identified which were responsible for the growth of the different parts of the fly. Their results had implications for higher organisms as well, including the embryonic development of humans.

After she had worked for years as a junior associate at the Max Planck Society in Tübingen, in 1985 Volhard became director of the Max Planck Institute of Developmental Biology. She and her associates continued working on *Drosophila*, and they also started to use zebrafish for their studies. They isolated their genes, identified them to learn which had particular functions, and studied their properties. They identified which genes determine, for example, the skull, the fins, and the scales of the fish.

In 2004, she started the Christiane Nüsslein-Volhard Foundation to give grants to young talented women with children, to facilitate childcare. She was more than willing to share her own experience when I asked whether she had ever experienced gender discrimination:

> Oh, yes, plenty! . . . First we can go back to when I was growing up. The general problem then was that women simply were not considered to be professionals—professional enough to have big important jobs. Therefore, they were just often overlooked and no one entrusted them with important

things. I have to say that my science was never discriminated against. So I had no problem whatsoever in getting my science recognized. But in the practical aspects, to get jobs, to get money, to get lab space, women have not been treated equally, I think.

At the time when I was a young scientist, men often had family and children and they got the better positions automatically. The professors always said, "but this is a man, he has to support and feed a family, so this is why he is going to get the job and you will not." This happened repeatedly to me. The same thing with promotions. They often said, you are a woman and there is a man and he deserves it more. Actually, my worst experience happened during my PhD; I collaborated with a man, I did most of the work, I wrote the paper and then he got first authorship because, as my boss said, "he has a family and for him this is important for his career. You are a woman and married, so it does not matter." This was particularly unjust because he gave up science right after his thesis and went to teach where he did not need the publication at all. Whereas, I did suffer in trying to find jobs later because I did not have this publication with my name as a first author. So discrimination started right away at the very beginning.

In the Max Planck Society, they created some jobs specifically for women and they raised the percentage of women, also at the independent group-leader level, dramatically in all their institutes, and filled these jobs in a short period of time. It turned out that all these women are very good in their jobs.[14]

In 1995, Christiane Nüsslein-Volhard was awarded the Nobel Prize in Physiology or Medicine, jointly with Edward B. Lewis (1918–2004) and Eric F. Wieschaus (b. 1947), "for their discoveries concerning the genetic control of early embryonic development." She did much of her prize-winning work in cooperation with Wieschaus. Lewis worked independently of them at the California Institute of Technology. One of her many other awards was the most prestigious Albert Lasker Award for Basic Medical Research, and another was the Gottfried Wilhelm Leibniz Prize, which is the highest honor given for research in Germany.

[14] M. Hargittai, "Christiane Nüsslein-Volhard: Biologist," in *Women Scientists: Reflections, Challenges, and Breaking Boundaries* (New York: Oxford University Press, 2015), 147–148.

Linda B. Buck

Linda B. Buck with another Nobelist, Aaron Ciechanover, during the Nobel festivities, 2004. Courtesy of Aaron Ciechanover.

Linda B. Buck (b. 1947) had a slow start, but when she found her goal in biology, her career took off spectacularly. A publication turned her attention to the sense of smell, and she eventually uncovered its foundation using the tools of molecular biology. Her discoveries were deemed worthy of the Nobel Prize.

Buck was born in Seattle, Washington, to an electrical engineer father of Irish roots and a homemaker mother, a daughter of Swedish immigrants. Buck was the second of three daughters. She grew up in a supportive environment; her mother told her never to settle for mediocrity. Buck studied psychology at the University of Washington with the idea in mind to become a psychotherapist, but by graduation time she had lost interest. For the next few years she was seeking a sustaining interest, and she finally found it in immunology. She enrolled for further studies in biology, still in her hometown.

She graduated in 1975 with a bachelor's degree. She was already 28 years old and started graduate school right away at the Department of Microbiology of the University of Texas at Dallas. She had good teachers, and her thesis advisor, Ellen Vitetta, pointed her toward a research career. For her degree, she investigated the functional properties of certain lymphocytes, immune cells made in bones. It fit nicely with her general interest in understanding the functioning of biological systems at the molecular level.

In 1980, Buck moved to Columbia University in New York for postdoctoral training. She knew she had to learn more about the modern techniques of molecular biology. She joined the laboratory of Richard Axel (b. 1946), which employed those tools in the study of memory. Their target was the nervous system of the sea snail *Aplysia*, whose relatively large nerve cells facilitated the experiments. This project originated with Eric Kandel (b. 1929), who had been Axel's mentor and who was also at Columbia. Axel and Buck were of the same age, yet he had become an established principal investigator when Buck was still looking for her own domain of science. She was seeking it relentlessly while enjoying independence and latitude, with the solid backing of Axel's laboratory behind her.

Buck's broad interests and search for a topic that would become the focus of her inquiry paid off. When she read a 1985 paper by Solomon H. Snyder and his group about the possible mechanism of odor detection, she felt that she had found what she was looking for. Olfaction, the sense of smell, relentlessly intrigued her. She had to find out how humans and other mammals are capable of detecting tens of thousands or even more odorous chemicals. How is it possible that they can distinguish between them? These were seemingly mundane but fundamental questions, and she was famous for asking good questions.

Her first step was to understand what happens in the nose where odorants are initially detected. There have to be receptors, molecules that have this capability, about whose existence people had made proposals, but nobody had identified them. Once she knew the questions to which she had to find the answers, her work became more defined. The receptors were proteins which she identified over the years with painstaking labor. The excellent relationship between Axel and Buck manifested in Axel's joining her project, and they co-authored a seminal paper in 1991. However technical it may have been, the title is remarkable for its clarity: "A Novel Multigene Family May Encode Odorant Receptors: A Molecular Basis for Odor Recognition."[15] This

[15] L. B. Buck and R. Axel, "A Novel Multigene Family May Encode Odorant Receptors: A Molecular Basis for Odor Recognition," *Cell* 65 (1991): 175–187.

paper has become the starting point for a number of other researchers who have entered this area of inquiry.

In 1991, Buck left Columbia University and became an assistant professor at the Department of Neurobiology of Harvard Medical School. This was an interesting situation. At 44, her age might have raised eyebrows for taking such a beginner position. On the other hand, those who decided to give her this position must have understood the significance of her discovery. Harvard neurobiology was a pioneering venue founded by the legendary Stephen W. Kuffler (1913–1980). Two of his disciples were awarded the Nobel Prize shortly after Kuffler's premature death. Buck established very good interactions with one of them, David Hubel (1926–2013). In 1994, Buck was named investigator of the Howard Hughes Medical Institute, which expanded her research possibilities considerably. She was fast rising on the academic ladder at Harvard, soon becoming associate and then full professor. In 1994, she met a biologist, Roger Brent, who became her lifelong partner; they married in 2006.

After a most successful decade at Harvard, in 2002 Buck returned to Seattle. She joined the Division of Basic Sciences at the Fred Hutchinson Cancer Research Center of the University of Washington. She was also appointed affiliate professor of physiology and biophysics. If geographically it is a big change from the East to the West Coast, two important features provided constancy in her professional life; one was the creative atmosphere of these research hubs, and the other was that she continued her focus on the mechanism of odor perception. This area of research includes the investigation of behavior under the influence of pheromones, thus it included a host of related projects. Pheromones are chemical substances released by mammals, insects, and other animals in order to affect the behavior and physiology of the other members of their species.

Richard Axel and Linda Buck received the 2004 Nobel Prize in Physiology or Medicine, each half of the prize and shared motivation: "For their discoveries of odorant receptors and the organization of the olfactory system." Her Nobel lecture reflected her happiness, and she ended with a pointed statement: "As a woman in science, I sincerely hope that my receiving a Nobel Prize will send a message to young women everywhere that the doors are open to them and that they should follow their dreams."[16] Buck certainly did follow hers.

[16] Linda B. Buck, "Biographical," Nobel Prize, April 5, 2021, https://www.nobelprize.org/prizes/medicine/2004/buck/biographical/.

Françoise Barré-Sinoussi

Françoise Barré-Sinoussi. Photograph by and courtesy of Marjolein Annegarn.

Françoise Barré-Sinoussi (b. 1947) was born and grew up in Paris, but she spent the summer holidays in the countryside of central France, where she learned to love nature. She became interested in science early on and chose a research career when she was so young that she could not fully understand what it meant. She enrolled at the science division of the University of Paris, but lacked motivation, and joined the Pasteur Institute to complete her studies. There, in laboratory work, something clicked. She continued her studies at the university but spent most of her time at the Institute. She became a full-time associate of the Pasteur Institute in the early 1970s, obtaining her PhD in 1974, before embarking on a postdoctoral internship at the U.S. NIH. Soon after her return to the Pasteur Institute, the AIDS epidemic broke out, 1981–1983. This area of research became Barré-Sinoussi's lifetime calling. She worked with the virologist Jean-Claud Chermann (b. 1939) as one of the first members of his research group. Luc A. Montagnier (b. 1932) was the administrative leader and was also very much involved in the research. They published a paper in 1983 about a virus associated with

AIDS, but they did not know what it was exactly; only later did it become known as HIV, the human immunodeficiency virus. The next year, in 1984, the U.S. scientist Robert C. Gallo and his colleagues at the NIH confirmed this discovery by Chermann and his colleagues.

In 1988, Barré-Sinoussi gained independence and established her own research group at the Pasteur Institute. As a continuation of her previous studies, she investigated various aspects of immune responses to viral infection. One of those many aspects was the mother-to-child transmission of HIV. She and her group published about 240 research papers, attended some 250 meetings, often giving presentations, and trained many young scientists. In 1992, she was named head of the Biology of Retroviruses Unit of the Pasteur Institute. HIV is also a retrovirus, and retroviruses are RNA viruses that insert a DNA copy of their genetic material into the host cell in order to replicate. Her unit was accordingly renamed in 2005 the Regulation and Retroviral Infections Unit. This laboratory has worked on vaccine research against HIV and has correlated the investigation of the possibilities of immunotherapy for protection against AIDS.

Barré-Sinoussi was active in various organizations dealing with the AIDS epidemic and with prevention. She recognized the importance of such work in developing countries, and the expression "resource-limited countries" often appears in the description of her activities. In 2006, she was elected a member of the Governing Council of the International AIDS Society, and served as president of the Society between 2012 and 2016. She has served in other capacities as well in the movement for preventing AIDS.

She received awards and prizes before her Nobel distinction: the King Faisal International Prize in the category of medicine in 1993, together with Chermann and Montagnier, and she was named an officer of the Legion of Honor in 2006. Following the Nobel Prize, she was named Commander Legion of Honor in 2009, and Grand Officer in 2013.

In 2008, Barré-Sinoussi and Montagnier shared half of the Nobel Prize in Physiology or Medicine for their discovery of HIV.[17] Nobody disputed that they deserved the distinction, but there was an uproar at the omission of Robert C. Gallo (b. 1937). Barré-Sinoussi and Montagnier isolated the virus and published their seminal finding in 1983 in *Science*. Alas, at that time they could not establish whether it was the virus causing AIDS. This

[17] The other half of this Nobel Prize went to Harald zur Hausen for his discovery of the link between human papilloma viruses and cervical cancer.

was accomplished in the next year by Gallo, whose group grew the virus, discovered its role, and opened the way for drug and vaccine development. Barré-Sinoussi and her colleagues readily recognized Gallo's contribution. In a letter to *Science,* a large number of well-known specialists, among them some members of the Karolinska Institute, whose Nobel Assembly was responsible for awarding the Nobel Prize in Physiology or Medicine, wrote, "Without Gallo's contributions, the relevance of this virus to AIDS might have not been recognized for years, and many thousands more lives would have been lost."[18] The title of the letter was "Unsung Hero Robert C. Gallo."

Conforming to French retirement rules, Françoise Barré-Sinoussi officially retired in 2015, but continued her activities for another two years before she fully retired in 2017. Her name is recorded in the annals of science and medicine as a hero in the fight against infectious deceases.

Barbara M. F. Pearse

The biologist Barbara M. F. Pearse (b. 1948) was born into an academic family; her father was a physics professor at Imperial College. She earned her bachelor's degree from University College London in 1969. She continued at UCL and earned a PhD based on her research on enzyme purification. At the Laboratory of Molecular Biology (LMB) of the Medical Research Council in Cambridge, UK, on an occasion in 1972, she put some of her samples into an electron microscope and saw a captivating pattern. Eventually she recognized the polyhedral shapes as resembling a European football.[19] The objects turned out to be coated vesicles, and Pearse was catapulted into an area of cell biology where she would make some of her most important discoveries.

Vesicles denote a variety of things in different areas of nature. In cell biology, a vesicle is a small volume of some liquid or some material within the cell membrane (except the cell nucleus) enveloped by a lipid bilayer

[18] Giovanni Abbadessa et al., "Unsung Hero Robert C. Gallo," *Science* 323, no. 5911 (2009): 206–207, https://science.sciencemag.org/content/323/5911/206.long.

[19] Today we call these shapes fullerenes after the buckminsterfullerene molecule (see "Elena G. Galpern," chapter 5). Such structures can be constructed from 12 pentagons and any number (except one) of hexagons.

Barbara M. F. Pearse in her laboratory at the MRC Laboratory of Molecular Biology, Cambridge, UK, 2000. Photograph by M. Hargittai.

within the cell or in the extracellular matrix. Vesicles may have important functions in metabolism and in the transport of materials, to mention just two examples in cell biology. Pearse decided to learn as much as possible about vesicles and their coating; she turned out to be uniquely qualified for this, given her accumulated experience in biochemistry. The coating consists of a large protein molecule whose pattern others had called a "basketwork." Pearse was looking for an appropriate name for the coated vesicles and found it in "clathrin." The term is of Greek/Latin origin and means "lattice-like." She published three seminal papers in 1975–1976, two solo and one with two distinguished LMB colleagues. It was especially appealing among her discoveries that she could correlate the intriguing polyhedral shapes of the clathrins with their function and mechanism of action. She opened a whole new area of inquiry, and an avalanche of reports followed her publications. Even people who had been skeptical as to the real existence of the coated vesicles and used to think of them as artifacts in sample manipulation, now joined the quest to uncover them and understand their role in cell biology. Pearse had the daunting task of staying abreast of the clathrin revolution she had started. She later reflected, "I had to paddle furiously to survive," but she did so, happily.[20]

Initially, Pearse was a visiting scientist at the LMB, and at some point transitioned to being a member of the scientific staff. Because her interest and work was moving toward structural research, she moved from the Protein and Nucleic Acid Chemistry Division into Structural Studies. Although she had also spent a visiting professorship at Stanford University—in itself, a great distinction—she always found sufficient stimulation at the LMB and never sought opportunities to do frontier science elsewhere.

LMB proved to be a friendly and accommodating environment for women scientists over the years. It is also a good place to stay informed of what is going on in science. In addition to its own greats, having had more Nobel laureates than some industrialized nations, there is a constant stream of visitors from all over the world, eager to learn and eager to share the advances in their fields.

Barbara Pearse is a representative of a number of successful women scientists who spent their entire careers at the LMB. It has also been the starting point in the careers of some of the other women scientists figured in this book, such as Elizabeth Blackburn (next entry) and Joan Steitz (chapter 5). They moved on to illustrious careers after brief stints as graduate students or postdoctoral fellows at the LMB.

Pearse's private life was also connected to the LMB: her husband, Mark Bretscher, has been one of its distinguished associates. Their daughter, Nicola, was born in 1987 and their son, Andrew, in 1981. Pearse's achievements were honored by her membership to the European Molecular Biology Organization in 1982 and by its Gold Medal in 1987. She was elected a Fellow of the Royal Society in 1988. Now in retirement, Pearse is enjoying her quiet life and gardening.

[20] Kathleen Weston, *Ahead of the Curve: Women Scientists at the MRC Laboratory of Molecular Biology* (Cambridge, UK: MRC LMB, 2020), 102.

Elizabeth H. Blackburn

Elizabeth H. Blackburn in Cambridge, UK, 2003, during the celebrations of the 50th anniversary of the discovery of the double-helix structure of DNA. Photograph by M. Hargittai.

Elizabeth H. Blackburn (b. 1948), a native of Australia, has spent her career in England and the United States. While at UC Berkeley, she made crucial discoveries in connection with telomeres and telomerase, the mechanism of protection of the DNA during cell division. For this work she was co-recipient of the 2009 Nobel Prize, together with her former student Carol W. Greider (next entry). Blackburn's activities have often been guided by social awareness and responsibility.

Blackburn was born in Hobart, Tasmania, in a family of two physicians. They soon moved to Launceston, Tasmania, where she attended school. When they moved to Melbourne, she completed her secondary schooling at the University High School. She enrolled at the University of Melbourne, majored in biochemistry, and earned bachelor's (1970) and master's (1972) degrees. She was a doctoral student of the two-time Nobel laureate Frederick Sanger (1918–2013) at Cambridge University, and received her PhD in 1975. Sanger's laboratory was at the LMB. He worked on the methodology of sequencing nucleic acids, that is, determining the order of nucleotides in them. For this, he received his second Nobel Prize in 1980. His first, in 1958, was for the methodology of sequencing proteins. In Cambridge, Blackburn married a fellow biochemist, John Sedat, and followed him to Yale University, where she completed her postdoctoral training. There, she worked with Joseph G. Gall (b. 1928), a cell biologist and renowned scholar of chromosome structure and function. He was well known for his ability to choose the most appropriate organism for investigating a specific question related to nucleic acids.

Blackburn embarked on the study of the single-cell organism *Tetrahymena thermophile*, especially suitable for learning about gene function, and noticed some peculiarities in the sequence of its DNA. In 1978, she left Yale and moved to the UC Berkeley. There, at the Department of Molecular Biology, Blackburn, her graduate student, Carol W. Greider, and another scientist, Jack W. Szostak (b. 1952), discovered something crucial to understanding how the genetic material chromosomes are copied during cell division without getting damaged or degraded. They found that the protection of the chromosomes is provided by so-called telomeres at the ends of the chromosomes. These telomeres are formed by an enzyme called telomerase. A telomere is a repeating nucleotide sequence added to the chromosome and is associated with some specific proteins. Telomerase protects the telomeres from degradation. At the risk of simplification, we may call it a double protection mechanism in which the telomeres protect the DNA of the chromosome and the enzyme telomerase protects the telomeres.

In 1990, Blackburn moved to another campus of the University of California, this time in San Francisco. She retired from UCSF in 2015 and became the president of the Salk Institute for Biological Studies in La Jolla, California. The Institute was founded in 1960 by Jonas Salk, famous for the Salk vaccine against infantile paralysis. Today, this international hub of biomedicine conducts research in three major areas: genetics, neuroscience, and plant biology. In neuroscience, it is ranked number one globally. Blackburn retired from the Salk Institute in 2017 but remained active beyond this second retirement.

Throughout her career, Blackburn has never shied away from controversial issues or from taking a stand, even if it meant defying the powers that be. In November 2001, President George W. Bush appointed an 18-member Council on Bioethics, and she was one of its members. She publicly criticized the motives and goals of the Council as they became clear to her, suggesting that its purpose was merely to lend support to the president's policy opposing stem-cell research and abortion. When, as a consequence, she was dismissed from the Council, there was widespread outrage among the scientific community. Stem-cell research was a hot topic, and it became a crossing point between politics and science. Blackburn's name reached a much broader public than it would have based solely on her scientific activities.

In 2009, Elizabeth H. Blackburn, Carol W. Greider, and Jack W. Szostak jointly received the Nobel Prize for Physiology or Medicine "for the discovery of how chromosomes are protected by telomeres and the enzyme telomerase." This distinction, of course, multiplied the strength of Blackburn's voice in societal matters. Her broad vision is reflected by her recent research interest: the impact of stress on telomeres and telomerase. In more practical terms, this is related to the consequences of psychological stress on ageing. For example, domestic abuse of women hinders their telomeres in cells and the enzyme telomerase, shortening the women's life and increasing morbidity.

Blackburn has an exceptionally long list of awards and honors, most of them from the time before her Nobel distinction.

Carol W. Greider

Carol W. Greider (b. 1961) was born in San Diego, California. Her father was a physics professor. The family moved to Davis, California, and she graduated from Davis Senior High School in 1979. Her bachelor's degree in biology (1983) was from the University of California, Santa Barbara. She spent some time doing research at the University of Göttingen in Germany during her undergraduate studies. Early on, she learned to overcome difficulties; she herself recognized her dyslexia before others noticed it, and did everything she could to turn this disadvantage into an advantage in her

Carol W. Greider, 2014. Photograph by Keith Weller/Johns Hopkins University School of Medicine. Source: Creative Commons.

research activities. In the process, she developed a sharper than usual sense of recognizing differences and viewing things from different angles, making choices that might appear unusual to others. Still, it was not easy.

She applied to 13 graduate schools and was accepted by two. Ironically, those two were among the top schools worldwide, the California Institute of Technology and the University of California, Berkeley. She chose Berkeley on account of the possibility of working with Elizabeth Blackburn. Greider joined her in November 1984 and completed her PhD in molecular biology in 1987. By the time Greider started her doctoral studies, Blackburn had discovered that there had to be an enzyme providing additional units of nucleotides (the telomeres) to the chromosomes as a protection during the process of cell division. Greider's principal discoveries concerned the enzyme telomerase and its action.

When Greider completed her studies with Blackburn, she embarked on a career that furthered this work on telomeres and telomerase. The ultimate goal was uncovering the relationship between telomeres and telomerase with disease and finding the appropriate remedies. From Berkeley, she moved first to the Cold Spring Harbor Laboratory in New York, and then to her latest venue, at least so far, the Johns Hopkins University School of Medicine. In 2006, Blackburn, Greider, and Jack W. Szostak received jointly the Albert Lasker Award for Basic Medical Research for the telomeres discoveries. As noted earlier, the same research trio was awarded the 2009 Nobel Prize in Physiology or Medicine "for the discovery of how chromosomes are protected by telomeres and the enzyme telomerase." Greider married a science historian in 1993. They had two children and divorced in 2011.

May-Britt Moser

The Nobel laureates May-Britt Moser and her partner in life and science, Edward I. Moser, are from Norway. Early on in their careers, they decided to understand more about the workings of the brain. Thanks to their success, we now know more about how grid cells are responsible for us being aware of our spatial position, the natural GPS system in our heads, as it were.

May-Britt Moser. Photograph by Torgrim Melhuus, Kavli Institute for Systems Neuroscience.

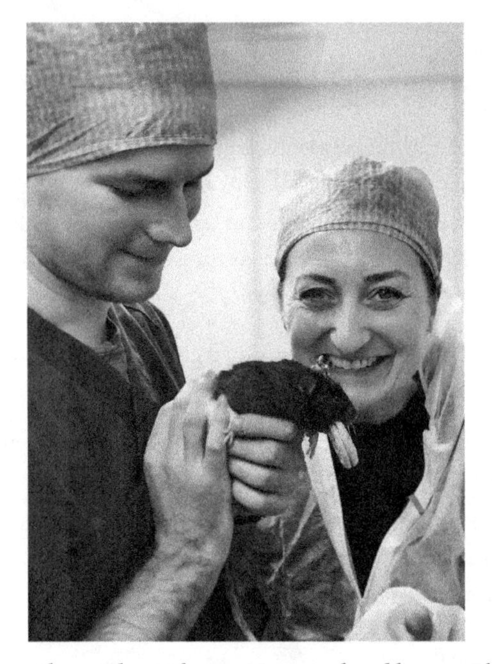

May-Britt Moser with postdoc Valentin Normand and his rat. Photograph by Rita Elmkvist-Nilsen, Kavli Institute for Systems Neuroscience.

May-Britt Moser (née Andreassen, b. 1963) was born in the small town of Fosnavåg in the county of Møre and Romsdal, in western Norway. Alesund and Molde are better known towns in the region, between Bergen and Trondheim. The family had a small farm, mainly taken care of by the homemaker mother; the father worked as a carpenter. May-Britt was the youngest among five children. She attended local schools and dreamed about becoming a doctor or a veterinarian. It was also her mother's childhood dream, and the profession of veterinarian attracted her because she loved animals. She wondered what made animals do what they do.

She did not stand out in school, yet her high school teachers did notice her because she knew unusual things. There was a physics teacher, very different from the typical male teachers in that he encouraged his female students to study and become engineers. She did not mind working hard, but what added to her drive was her mother's warning that if she did not work hard, she would have to become a housewife. After high school she enrolled at the University of Oslo and enjoyed it. She was far from knowing yet what she liked best, except that she tended toward biology and geology, but she loved mathematics and physics as well. She met Edward Moser, whom she had known vaguely in high school, and they started making plans for their further studies together.

They decided on psychology, with an eye to learning as much as possible about how the brain works. Although psychology was not the best recommendation for becoming a researcher in experimental neuroscience, a neuroscience professor, Per Andersen, took them on for their master's thesis work. They continued their psychology studies during the day and immersed themselves in neuroscience experiments on evenings and weekends. Their experiments were about signal transmission between the neurons—nerve cells—in the brains of rats. May-Britt became a psychologist in 1990 at the Department of Psychology and in 1995 earned her PhD in neurophysiology at the Faculty of Medicine. Gaining the necessary funding for two PhD projects, hers and his, was not easy, but they succeeded. Referring to the targeted doctoral mentor, she wrote, "I kept going to his office with the application I wrote to see if he had changed his mind, and finally he just gave in—he always had to give in, like my dad typically gave in when I insisted."[21]

May-Britt and Edward married in 1985 (they divorced in 2016, but have continued their joint research work). They had two children, Isabel in 1991 and Ailin in 1995, but this did not slow down their ambitions. She regularly took their daughters to the laboratory, or both parents shared the responsibility of staying with the girls at home. She took them to scientific meetings and did not mind breastfeeding them in public. Already during their PhD work they started building international connections and spent some time in the laboratories of renowned scientists in their field. One was Richard Morris (b. 1948) at the University of Edinburgh; another was John O'Keefe (b. 1939) at University College London. O'Keefe was the great pioneer in the field for having discovered the place cells in the brain, which produces signals when the animal is at a specific place. Both Morris and O'Keefe selflessly and enthusiastically helped the Mosers in gaining the necessary experience.

When they completed all their studies in Oslo, the Mosers decided to continue their careers in Trondheim at the Norwegian University of Science and Technology (NUST, what used to be the Norwegian Technical University). Again they had to overcome the hurdle of needing two positions rather than one, and again they succeeded. They were not the easiest of employees. Rather

[21] May-Britt Moser, "Biographical," Nobel Prize, April 4, 2021, https://www.nobelprize.org/prizes/medicine/2014/may-britt-moser/biographical/.

than being happy to teach and earn a salary (two salaries, to be exact), they insisted on establishing a laboratory with all the necessary equipment for brain research. Apparently their plans convinced granting institutions. The European Commission gave them funding for three years in 2000, and the Norwegian Research Council for 10 years in the framework of their Center of Excellence Program. This became known as the Center for the Biology of Memory at NUST, and became an international hub of their science. The support made it possible to bring in the best researchers for substantial visits. In 2002, they already had a seminal publication about the place cells in the hippocampus of the brain. This was a substantial step on the path toward their milestone discovery of the grid cells a few years later, in 2005. A grid cell is a neuron, a nerve cell, that is activated at regular intervals allowing the animal (or human) to understand its position in space.

In 2007, the Mosers' laboratory was selected to be a Kavli neuroscience institute by the Kavli Foundation. It enhanced their prestige, and more important, it ensured funding for their basic research in perpetuity. The scientists reciprocated by producing new discoveries and groundbreaking results. Neuroscience has become recognized as a "Norwegian science." In 2011, the Norwegian government decided to fund the Norwegian Brain Initiative. This made it possible for the Mosers to establish the Norwegian Brain Center, which opened in 2012 in partnership with other institutions in the country. Also in 2012, with the expiration of the first 10-year-long support for the Center of Excellence, there was an award of a second 10-year-long support by the Norwegian Research Council. This is considered to be a new Center for Neural Computation, and May-Britt Moser was appointed its inaugural director. The research has expanded its purview; the goal now is to understand how the brain generates a universal map of the environment.

The Mosers received a number of awards and prizes, both shared and simultaneous. An exception was the Erna Hamburger Prize in 2016, which was given only to May-Britt and which "distinguishes influential, leading women scientists who are transforming their field and executing change." John O'Keefe received half of the Nobel Prize in Physiology or Medicine for 2014, and the other half was divided, a quarter each, to the two Mosers "for their discoveries of cells that constitute a positioning system in the brain." Their determination to understand more about the workings of the brain continues. Their two daughters refer to their parents' lab as their third child.

7
Physicians, Surgeons, and Nurses

There are few women surgeons, but women doctors are not scarce. However, this was not always so; women fought a hard uphill battle for the possibility of a career in medicine.

Elizabeth Blackwell

Elizabeth Blackwell, by Alexis Delacoux. Source: Wellcome Collection.

Elizabeth Blackwell (1821-1910) was the first woman to receive a medical degree in the United States. She was born in Bristol, England. Her family moved to the United States in 1832, where her career began with a variety of teaching jobs. In 1847, she embarked on a medical degree at the Geneva Medical College (now the Hobart and William Smith College in Geneva, New York) and graduated in 1849. In the same year, she returned to Europe and continued her studies in France. She hoped to be accepted as a physician in training, as France was considered more progressive than Britain at the time with respect to the inclusion of women in the profession. She wanted to become an obstetrician, though the institution that accepted her as a student insisted on treating her as a midwife rather than a physician. Eventually she moved to London and received further training at St. Bartholomew's Hospital.

In 1851, she moved back to America and established her practice in New York City. Two years later, along with two female colleagues, she established the New York Infirmary for Indigent Women and Children. In 1858, she again returned to England, where a new medical act recognized doctors with foreign degrees, and she entered the medical register in 1859. She mentored Elizabeth Garrett Anderson (see below) and others in their medical studies, corresponded with Lady Byron—wife of the famous poet—about women's rights, and befriended Florence Nightingale (see below). There are beautiful memorials of Blackwell in Asheville, North Carolina, and on the campus of the former Geneva Medical College, the school that admitted her in 1847 after she had been refused by 16 others.

Rebecca Lee Crumpler

Rebecca Lee Crumpler (née Davis, 1831-1895) was the first African American medical doctor in the United States. She was born in Delaware and grew up in Pennsylvania, where she was raised by her aunt, who was a sort of substitute doctor in their neighborhood. Rebecca went to a quality private high school and for years worked as a nurse. In 1860, she was accepted by the New England Female Medical College, where she was the only African American student. She graduated in 1864 and started working in Boston. Her studies were interrupted by the Civil War, and when it ended the Crumplers moved to Richmond, Virginia. Interesting to note that her

Rebecca Lee Crumpler. Source: Wikipedia.

alma mater was founded in 1848 by two physicians, Dr. Israel Tisdale Talbot and Dr. Samuel Gregory, and the instruction started in 1850. The school closed in 1873, and Crumpler remains its only African American graduate.

An African American doctor was a great rarity at the time. In 1860, there were only 300 women out of 54,543 physicians in the United States, and none was African American.[1] Crumpler worked under a program of the Freedman's Bureau of the State of Virginia, whose task was assisting millions of former slaves to make the transition to freedom. She excelled in treating sick women and children. In 1869, the Crumplers moved to Boston and she worked in an African American neighborhood; in 1880, they moved to Hyde

[1] Celebrating Rebecca Lee Crumpler, first African-American woman physician | PBS NewsHour. Howard Markel.

Park, New York. It is supposed that she continued her activities during the rest of her life, but little information is available. She died in Boston. Her *Book of Medical Discourses in Two Parts* (1883) has recently been republished.[2]

Rebecca Cole

Rebecca Cole. Artist unknown. Source: Wikipedia.

Rebecca Cole (1846–1922) was the second African American female doctor. She was born in Philadelphia. She went to a good high school for people of color and received quality training in math and Latin and Greek, in addition to other subjects. She graduated in 1863 and enrolled at the Woman's Medical College of Pennsylvania, founded by Quaker abolitionists and

[2] Rebecca Crumpler, *A Book of Medical Discourses: In Two Parts* (Classic Reprint) paperback. Forgotten Books, 2018.

temperance reformers in 1850. At the time, it was called the Female Medical College of Pennsylvania, and it was among the world's first medical schools for women. Upon her graduation in 1867, she went to work at Elizabeth Blackwell's New York Infirmary for Indigent Women and Children. Her tasks included visiting impoverished tenements and teaching prenatal care and hygiene. For a while she practiced in South Carolina, then returned to Philadelphia and in 1873 helped to establish a Women's Directory Center. Its purpose was to provide legal services to destitute women and children. Cole spent the rest of her career providing these services. At the time of her death, there were still only 65 African American women doctors in the United States.

Elizabeth Garrett Anderson

Elizabeth Garrett Anderson (née Garrett, 1836-1917) was the first English woman medical doctor. She was born in East London. Elizabeth Blackwell convinced the younger Elizabeth to become a doctor. She studied to become a nurse, but attended classes for male medical doctors until she was barred from doing so on account of the protesting male students. However, she received a certificate from the Society of Apothecaries in 1865, as women were not explicitly excluded by the rules of the Society. Soon afterward, the Society

Memorial plaque of Elizabeth Garrett Anderson at 20 Upper Berkley Street, London W1. Photograph by M. Hargittai.

Elizabeth Garrett Anderson. Source: Wellcome Collection.

changed its regulations to prevent other women from using this path to becoming medical professionals.

With her certificate, Garrett opened a dispensary for women in London, became a visiting physician to the East London Hospital in 1870, and founded the New Hospital for Women in London in 1872. It was staffed exclusively by women doctors; today, it is the Elizabeth Garrett Anderson Hospital for Women. In 1871 she married James Anderson and they soon had three children. She continued her career in medicine as the legislative environment was slowly improving for women entering the medical profession, in part due to her example. In 1873, she was accepted as a member of the British Medical Association and remained its only female member for the next 19 years. She founded the London School of Medicine for Women, and in 1883 she was appointed its dean.

The blue plaque shown here was erected on the façade of the house where Garrett Anderson lived after she married. There is another plaque (not shown) at the site where her childhood house stood; today the Metropolitan University stands there, 41-47 Commercial Road, E1. There is the Elizabeth Garrett Anderson Wing of the UCL Hospital and an Elizabeth Garrett Anderson Gallery at 130 Euston Road, NW1.

Louisa Brandbeth Aldrich-Blake

The memorial of Louisa Brandbeth Aldrich-Blake in Tavistock Square. Photograph by M. Hargittai.

The memorial of Louisa Brandbeth Aldrich-Blake was designed by Edwin Lutyens and consists of two identical busts. Photograph by M. Hargittai.

Louisa Brandbeth Aldrich-Blake (1865–1925) lived decades later than Elizabeth Blackwell, yet she was still among the first British women in medicine. She graduated from the Royal Free Hospital's School of Medicine for Women in 1894. She continued her studies for a master's degree in surgery, thereby becoming the first British woman surgeon. She was a pioneer in treating cervical and rectal cancers and had publications in this area. She worked at her alma mater, devoted much attention to training her students, and in 1914 was made dean of the School of Medicine for Women. She took an active role in the British Medical Association. The careers of Blackwell, Garrett Anderson, and Aldrich-Blake contributed greatly to advancing the opportunities for women in medicine, much beyond forging their own advancement. Looking back, the pace of progress may appear slow, but from their perspective, they took giant steps ahead in their own lifetimes.

Florence R. Sabin

Florence R. Sabin (1871–1953) was born in Central City, Colorado, to a mining engineer father and a schoolteacher mother. When Florence was 7 years old her mother died of puerperal fever. Still she and her older sister, Mary, received a good education thanks to the efforts of her father and one of her uncles. She attended Smith College, majored in zoology, and decided on a career in medicine. She graduated Smith in 1893 and taught high school for three years to save money for further studies. In 1896, she enrolled at the

Florence Sabin as a department head at the Rockefeller Institute.
Source: Archives of the U.S. National Library of Medicine.

Statue of Florence R. Sabin by Joy Buba, 1959, in the National Statuary Hall Collection at the U.S. Capitol in Washington, D.C. Photograph by Istvan Hargittai.

Johns Hopkins Medical School, where she came under the influence of an excellent mentor, Franklin P. Mall, who recognized her talent and encouraged her toward a career in research.

Sabin was among the first women MDs graduating from Johns Hopkins University. Following graduation, she obtained an internship from this school and thus could continue her research there. She chose embryology for her investigations, staying in close contact with Mall. The influential physician William Osler, one of the founding professors of Johns Hopkins Hospital, was impressed by her progress and dedication and helped her to stay on at Johns Hopkins. She was rising on the academic ladder and, in 1917, was the first female full professor in an American medical college with the title of professor of histology.

Another first occurred in 1924, when she was elected as the first female president of the American Association of Anatomists. In 1925, she was elected a member of the National Academy of Sciences of the U.S.A., the first woman scientist bestowed such an honor. This was an exceptional recognition considering that, for example, the Royal Society (London) elected its first woman members only 20 years later. Another important event in her life happened in the same year, 1925, when she was appointed head of the Department of Cellular Studies at the Rockefeller Institute (as it was then) of Medical Research. She was the first woman appointed full member of the Institute. Her main research was in the pathology of tuberculosis. She studied the responses of the immune system to the tuberculosis bacteria.

Sabin retired from the Rockefeller Institute in 1938 and moved back to Colorado. She did a great deal to improve the state's public health system. She and her efforts were not always popular among politicians, who were not eager to spend money on her projects, which benefited a broad segment of the population. To set an example, she donated her salary to the cause for a period of three years. She succeeded in reducing the number of victims of tuberculosis. When she was offered an office to direct the state's public health efforts, she realized that the politicians expected nothing from an old lady, which is why they had appointed her. Instead, she had the opposite effect, helping to defeat politicians who opposed healthcare reform. For her work she was awarded the prestigious Albert Lasker Public Service Award in 1951. This was also the year of her second and final retirement.

The National Statuary Hall Collection at the U.S. Capitol in Washington, D.C. is a special venue for memorialization. There are 100 statues there; every state has contributed two statues honoring notable persons in their history. Colorado chose to honor Florence R. Sabin.[3] Her statue stands in the Hall of Columns, not part of the usual route for tourists. It depicts a lifelike Sabin with a book and a microscope. She is one of only nine female historical figures among the 100 honorees.

The Moscow Higher Courses for Women

The Pirogov Medical University, one of the two main medical schools in Moscow (the other is the Sechenov Medical University), traces its origin to the Higher Courses for Women. The first lectures in medicine at these Higher Courses were delivered in 1906, and the first women physicians graduated in 1912. It was at this time that the main building of the Higher Courses was completed. Later the building belonged to the Moscow Pedagogical University.

[3] The other Colorado honoree, the former astronaut John L. Swigert Jr. (1931–1982), received his statue in 1997; it stands in the Capitol Visitor Center.

The former main building of the Higher Courses for Women in Moscow stands at the intersection of Malaya Pirogovskaya, Rossolino, and Kholzunov Streets. It was completed in 1912. Photograph by M. Hargittai.

The reliefs on its façade symbolize the broad scope of the Higher Courses for Women. Photographs by M. Hargittai.

The Moscow Higher Courses for Women was initiated in 1863 by Vladimir I. Gerie (Guerrier, 1837–1919), a professor of history at Moscow University. He found it unjust that the ministerial order rejected the possibility of women enrolling at Moscow University. He had had excellent experience with female students who were dedicated and well-prepared but could only become private tutors. He decided to establish an institution of higher education for women. It opened in 1872 as a private school, for which he served as its first director.

In 1911, the authorities blatantly violated the autonomy of Moscow University, and many of its professors resigned in protest. Some of the most prominent professors joined the Higher Courses for Women and thus enhanced greatly its academic strength. However, the Courses were short lived; following the Bolshevik Revolution all institutions of higher education became coeducational. In 1918, the Higher Courses was transformed into the Second Moscow University and, eventually, into the Pirogov Medical University. There were other higher courses for women in Russia established in the final decades of the 19th century, and they all operated only until the victory of the Bolshevik Revolution.

A few students of the Moscow Higher Courses are mentioned here. **Ekaterina A. Kost** (1888–1975) graduated in 1913 and had a brilliant career in medicine. She rose to the rank of professor. Hematology was her main specialty, but her monographs covered a wide range of topics in medicine. **Nadezhda N. Sushkina** (1889–1975) was a microbiologist and a noted specialist in soil science. She was a pioneer in evaluating the role of microorganisms in natural formations. **Lidia K. Lepin** (or Leipina, 1891–1985) was a physical chemist, one of the first women in the Soviet Union who received a doctorate in chemistry. She was a member of the Latvian Academy of Sciences. **Pelageya Ya. Kochina** (née Polutbarinova, 1899–1999) completed only part of her studies at the Higher Courses before it was reorganized. She specialized in hydrodynamics, and beginning in 1970 she headed the mathematical section of the Institute of Problems of Mechanics in Moscow.

The Moscow Higher Courses for Women received the right to issue diplomas (master's degree equivalent) starting in the academic year 1915–1916. In 1900, it had 223 students; in 1918, its last year of operation, 8,300 students attended, second only to Moscow University.

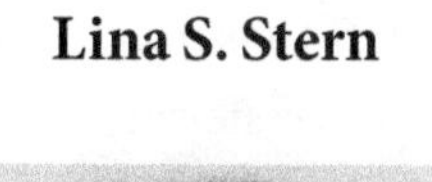

Lina S. Stern

Lina S. Stern, 1910s. Source: U.S. National Library of Medicine, Bethesda, Maryland.

Lina S. Stern's grave memorial in the Novodeviche Cemetery, Moscow. Photograph by M. Hargittai.

Lina Solomonovna Stern (1878–1968) was a biochemist-physiologist. She was born in Liepāja (today, in westernmost Latvia; then, in the Russian Empire). She was among the first women to study at the University of Geneva, Switzerland. After graduation she stayed in Geneva and carried out research in biochemistry and neurology, becoming the first female professor at the university. At the invitation of the Soviet government, she moved to Russia in 1925, where she was appointed head of a research laboratory in a medical school in Moscow. In 1929, she became the first director of the Institute of Physiology of the Soviet Academy of Sciences. She was successful in her research; her most important result was in the study of what is today called the blood-brain barrier. She was much valued both in the young Soviet state and in the West. She was elected to the Leopoldina Academy in Germany and in 1939 became a full member of the Soviet Academy of Sciences—the pinnacle of a scientist's career in Russia—the first woman upon whom such an honor was bestowed.

Also in 1939, she joined the Communist Party, and in 1943 received the Stalin Prize. During World War II, many of her research achievements were turned to practical uses. She introduced novel treatments for neurological disorders, and her procedures saved thousands of lives on the front. After the war, her career fell victim to Stalin's anti-Semitism as well as his distrust of science and scientists, a period that lasted until his death in 1953. First, she was stripped of all her positions. She was a member of the Jewish Anti-Fascist Committee and of the Women Antifascist Committee. The former was an organization originally set up by the Soviet government to mobilize Jews for the worldwide struggle against Nazi Germany. At one point, all members of the Jewish Antifascist Committee were arrested, tried, and—with the exception of Stern—executed. Stern was incarcerated instead and later sent into internal exile. She could return to Moscow only after Stalin's death, when she was exonerated and her membership in the Academy of Sciences was reinstated. She continued her research activities and headed the Department of Physiology at the Biophysics Institute until her death in 1968.

Women in Medicine in London

London is rich in memorials to scientists, but rare are those honoring women. On a recent visit, my husband and I sought out the memorials to women in medicine as part of our broader explorations of memorials to scientists.[4]

[4] I. Hargittai and M. Hargittai, *Science in London: A Guide to Memorials* (Springer Nature, 2021).

Annie McCall. Photograph-portrait by Deneluain. Source: Wellcome Collection.

Mosaic-portrait of Annie McCall on the wall of Morley College, 61 Westminster Bridge Road, London SE1. Photograph by M. Hargittai.

Jane Harriet Walker. Source: Wellcome Collection.

Annie McCall (1859-1949) was a native of Manchester but studied at several European universities—in Göttingen, Paris, Bern, and Vienna. She continued at the London School of Medicine for Women, the first British medical institution to train women to become doctors. McCall, who specialized in midwifery, was among its first 50 graduates. She founded the Clapham Maternity Hospital, which was later renamed the Annie McCall Maternity Hospital.

Jane Harriet Walker (1859-1938) specialized in treating tuberculosis. She adopted an approach, initiated in Germany, whereby the patients were encouraged to spend as much time as they could in the open air rather than in stuffy, warm, indoor environments. She opened sanatoria for treating tuberculosis patients and was a founding member and the first president of the Medical Women's Federation. She was also the first female member of the Council of the Royal Society of Medicine.

Portrait of Elsie Inglis by Frances Balfour. Source: Wellcome Collection.

Bust of Elsie Inglis by Ivan Mestorovic, 1918, at the Imperial War Museum, London. Photograph by M. Hargittai.

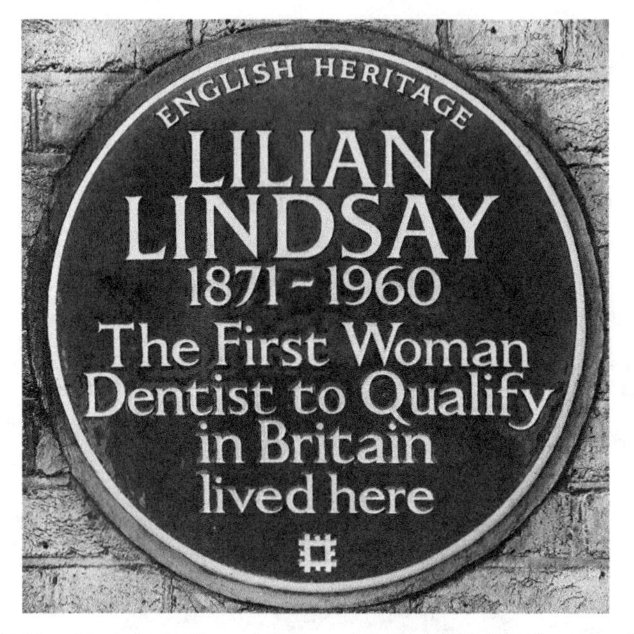

Plaque of Lilian Lindsay, 23 Russell Square, London WC1H. Photograph by and courtesy of Steve Roffey.

Elsie Inglis (1864–1917) had an enlightened father who supported her determination to study medicine at a time when this was not an accepted career for women. Later in her life, in order to help others, she and her father established the Edinburgh College of Medicine for Women. After gaining her medical qualifications, she worked for some time at the New Hospital for Women pioneered by Elizabeth Garrett Anderson. When the University of Edinburgh opened its medical training to women, Inglis obtained her Bachelor of Medicine/Bachelor of Surgery degree in 1899. When in World War I she offered her services to the British War Office, she was rather impolitely refused. In contrast, the French government was happy to take up her offer and gave her and her team an assignment in Serbia. They treated wounded Allied soldiers and helped improve hygiene, which reduced the spread of typhus and other infectious diseases. She tried to organize similar efforts for the Russian front, but cancer prevented her from further actions. In addition to British recognition, she was awarded the highest order of Serbia, where she has become a legend.

Lilian Lindsay (née Murray, 1871–1960) always wanted to become a dentist. She could not study in London because she was a woman, so she moved to Edinburgh, where she could. Even there she sometimes experienced hostility, accused of taking the place of a man who would be the breadwinner for his family. Upon graduating from the Edinburgh Dental Hospital and School in 1895, she became the first woman in the United Kingdom to qualify as a dentist. Later the same year, she was the first woman to join the British Dental Association (BDA), then, half a century later, the first woman president of the BDA. Other firsts preceded and followed this remarkable achievement. She was active in establishing and enriching the BDA library and museum, and she published a book and many articles related to the history of dentistry in Britain. When, in 1920, her husband, Robert Lindsay, was appointed to a position in the BDA, they moved to London and lived in an apartment above its headquarters on Russell Square. Her memorial plaque is now on the façade of this building.

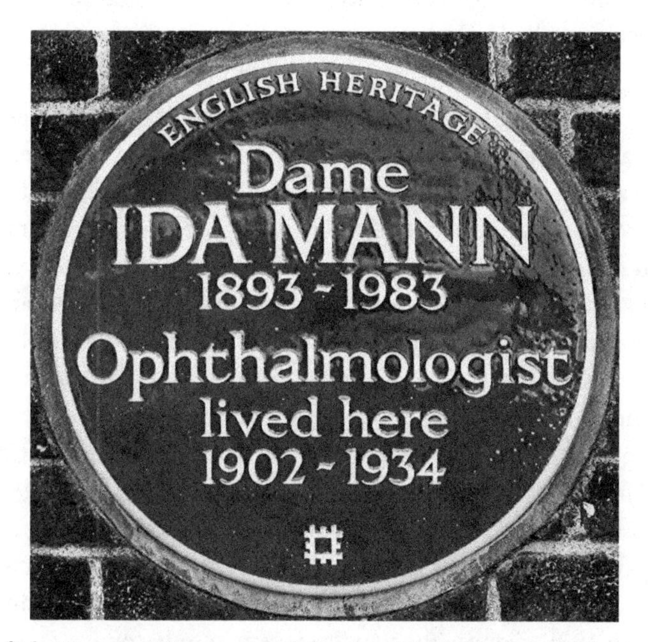

Plaque of Ida Mann, 13 Minster Road, London NW2. Photograph by and courtesy of Steve Roffey.

Plaque of Dame Sheila Sherlock, 41 York Terrace East, Westminster, London NW1. Photograph by M. Hargittai.

Ida Mann (1893–1983) was an ophthalmologist whose distinguished career started at a time when women seldom could embark on such a career. By 1927 she had a staff position at the Moorefield Eye Hospital and a private practice in Harley Street. Her principal research concerned embryology and the genetic and social factors influencing the early development of the eye. She was involved with chemical defense during World War II. After the war, she and her husband moved to Australia, where she did much work for public health.

Dame Sheila Sherlock (1918–2001) was declined repeatedly by several English medical schools in 1935–1936, but was accepted by the University of Edinburgh. She graduated in 1941 at the top of her class and attended the Royal Postgraduate Medical School, Hammersmith Hospital, in London (today, part of the Imperial College School of Medicine). Focusing her research on hepatitis, she became an internationally renowned authority on the liver. In 1951, she was the youngest woman to become a Fellow of the Royal College of Physicians. In 1959, she was appointed professor of medicine at the Royal Free Hospital School of Medicine in London, the first female professor of medicine in the United Kingdom.

Painting of Margaret Turner-Warwick by David Poole, 1992, Royal College of Physicians, London. Photograph by M. Hargittai.

Margaret Turner-Warwick (1924–2017) was a physician, most famous for her work on respiratory diseases. In her youth, she suffered from tuberculosis; thereafter her interest settled on lung function, cystic fibrosis, and asthma. Her recommendation of treatment with corticosteroids has been broadly followed. Among other contributions, she called attention to the dangers of working with asbestos.

Bust of Dame Cicely Saunders by Shenda Amery, 2002, in the Central Hall of St. Thomas' Hospital, London. Photograph by M. Hargittai.

Plaque of Melanie Klein, 42 Clifton Hill, London NW8. Photograph by and courtesy of Steve Roffey.

Plaque of Anna Freud, 20 Maresfield Gardens, London NW3. Photograph by and courtesy of Steve Roffey.

Dame Cicely Saunders (1918–2005) was the founder of the modern hospice movement. She was trained as a medical social worker and, in 1957, qualified as a doctor, which added weight to her teachings. She researched the issue of terminal care and was a prolific author on the subject. In 1989, she was appointed to the Order of Merit, the pinnacle of the British honors system.

Melanie Klein (née Reizes, 1882–1960) and **Anna Freud** (1895–1982), Sigmund Freud's daughter, were Austrian-born British contributors to and pioneers of child psychoanalysis.

The couple Michael Balint (1896–1970) and **Enid Balint** (1903–1994) worked as psychoanalysts. Together with a group of general practitioners, they conducted a large study in the 1950s of the doctor-patient relationship. Their recommendations improved the process of diagnosis and treatment of patients by general practitioners. To mark the 50th anniversary of the formation of the Balint Society in 1969, and honoring the Balints' work,

Plaque of Michael Balint and Enid Balint, 7 Park Square West, London NW1. Photograph by M. Hargittai.

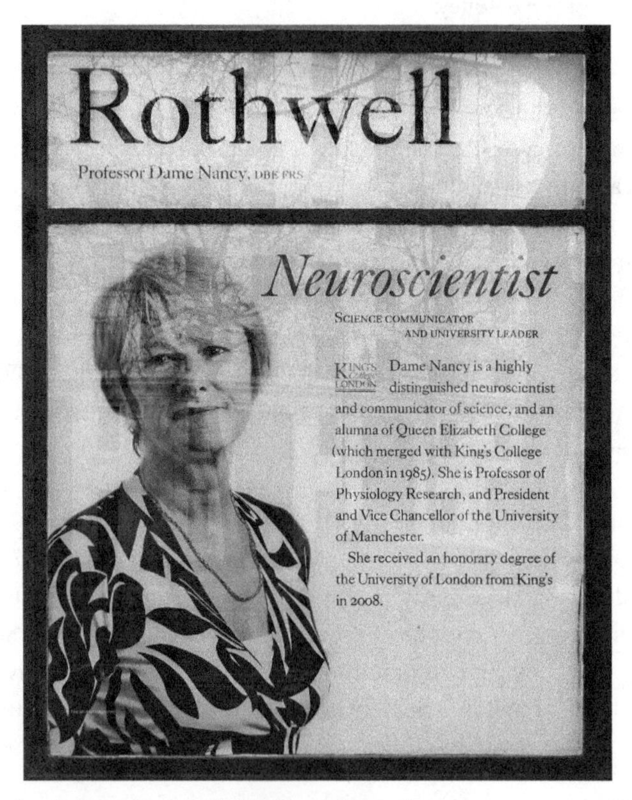

Photograph of Nancy Rothwell displayed at the downtown campus of King's College, London. Photograph by M. Hargittai.

a memorial plaque was unveiled on the façade of the building where they lived and worked.

Michael Balint was a graduate of what is today the Semmelweis University in Budapest and studied psychoanalysis with Sándor Ferenczi. His first wife, **Alice Balint** (1898–1939), was an outstanding psychoanalyst in her own right. The autocratic and anti-Semitic Horthy regime in Hungary did not tolerate the Balints' science, so they fled to the United Kingdom in 1938. Alice soon succumbed to a devastating illness. Balint met Enid at the Tavistock Institute, and they became partners in work and, eventually, spouses.

Nancy Rothwell (b. 1955), a physiologist, has held leadership positions both in academia and in the pharmaceutical industry. She studied at the University of London and in 1987 was awarded a Doctor of Science degree by King's College. Her research interests have included obesity and involuntary weight loss, the role of inflammation in brain disease, and the treatment of stroke.

Nurses

Statue of Florence Nightingale, "The Lady of the Lamp," by Arthur George Walker (1915), part of the Crimean War Memorial, Waterloo Place, London SW1. Photograph by M. Hargittai.

Memorial plaque honoring Florence Nightingale, 10 South Street, London W1K. Photograph by and courtesy of Steve Roffey.

Photograph of Florence Nightingale by Kilburn. Source: Wellcome Collection.

The nurse **Florence Nightingale** (1820–1910) is an iconic figure in British medical and military history. She revolutionized the treatment of the wounded in the Crimean War (1853–1856) and the functioning of hospitals. She not only practiced her profession but recognized the need to record the proper ways of operating institutions. In her *Notes on Hospitals* (1863), she was not afraid to state a seemingly obvious expectation, which is still not that trivial today considering the common incidence of hospital infections: "It may seem a strange principle to enunciate as the very first requirement in a Hospital that it should do the sick no harm."[5]

Four bas relief panels decorate the plinth of her statue at Waterloo Place, depicting the following scenes: "Caring for the injured"; "Negotiating with military and political leaders"; "Challenging medical and hospital managers"; and "Teaching and inspiring nurses." Nightingale assisted in the founding of the nursing school at St. Thomas' Hospital, which was revolutionary in that it was a secular institution, and it was the first such nursing school worldwide. The graduates are known as Nightingales. There is a copy of her statue at the Central Hall of St. Thomas' Hospital. There is also a Florence Nightingale Museum on the campus of the Hospital at 2 Lambeth Palace Road, London, SE1.

Mary Seacole's mosaic portrait on the wall of Morley College, 61 Westminster Bridge Road, London SE1. Photograph by M. Hargittai.

[5] W. F. Bynum and Roy Porter, eds., *Oxford Dictionary of Scientific Quotations* (Oxford: Oxford University Press, 2005), 465.

Plaque honoring Mary Seacole, 14 Soho Square, London W1. Photograph by M. Hargittai.

Statue of Mary Seacole by Martin Jennings, 2016, in front of St. Thomas' Hospital, London. Photograph by M. Hargittai.

Steel statue of Mary Seacole at the southern end of St. Mary's Terrace, London W2. Photograph by M. Hargittai.

Mary Seacole (1805-1881) was a British Jamaican nurse who demonstrated self-sacrifice and dedication during the Crimean War. Although her services were declined by the authorities, she still attended to and comforted the wounded and applied Jamaican and West African remedies to those in her care. She received much recognition posthumously; hers was the first memorial to a Black woman in the UK.

Bust of Theodora Turner by Robert Dawson, 2002, in the Central Hall of St. Thomas' Hospital, London. Photograph by M. Hargittai.

Theodora Turner (1907–1999) was trained as a nurse and performed her duties both in peacetime and in war. In World War II, she was there at the evacuation of the British troops in Dunkirk, and she participated in rebuilding St. Thomas' Hospital after it was struck by 13 German bombs. At the end of her career, she was the matron of nurses at St. Thomas' Hospital.

8
Inventors and Technologists

In addition to the category of the physicist, that of the inventor and technologist is perhaps the most difficult for women to access given that invention has for so long been seen as the natural domain of men. However slowly, this perception is changing.

Hertha Ayrton

Hertha Ayrton (1854–1923) was a British physicist, engineer, and inventor. She was born Phoebe Sarah Marks. Her father was a Polish Jewish immigrant who died early. Poverty made her childhood difficult, but her talent for science and mathematics showed, so she received support from family and friends and attended Girton College at Cambridge University. While still a student, she constructed an apparatus for measuring blood pressure. In 1880, she passed the necessary exams to study for a degree at Cambridge University, which did not award full degrees to women at the time, so she moved to London. She received her Bachelor of Science degree in 1881 from

Hertha Ayrton. Source: Wikipedia.

Hertha Ayrton's memorial plaque, 41 Norfolk Square, London W2. Photograph by M. Hargittai.

University College London. In 1884, she filed for her first patent, a line-divider, a mechanical instrument capable of dividing a line into any number of equal parts and enlarging and reducing figures. She successfully registered 25 more patents during her life. Most of her inventions and patents concerned arc lamps and electrodes. This was after she had extended her studies to electrical engineering. In 1885, she married her former UCL professor of electricity, William Edward Ayrton.

In 1900, she was nominated as a Fellow of the Royal Society. There had not been any female FRS until then, and she was not to be the first. The denial was based on a curious reasoning: the committee observed that her husband was already a Fellow. Because under British law husband and wife were considered to be one person, she could not be a Fellow because she did not represent a separate person.[1] Though shut out of its fellowship, she did become the first woman awarded a prize by the Royal Society, receiving the Hughes Medal in 1906 for her achievements in studying the electric arc and the motion of ripples in sand and water. She was active in politics and fought for strengthening women's positions in scientific societies and elsewhere. In 2010, a panel of experts of the Royal Society named her one of the 10 most influential women scientists in British history, alongside Caroline Herschel (astronomer), Mary Somerville (physicist), Mary Anning (paleontologist), Elizabeth Garrett Anderson (physician), Kathleen Lonsdale (crystallographer), Elsie Widdowson (nutritionist), Dorothy Hodgkin (crystallographer), Rosalind Franklin (biophysicist), and Anne McLaren (geneticist).

[1] The crystallographer Kathleen Lonsdale and biochemist Marjory Stephenson became the first female Fellows, in 1945.

Katharine Blodgett

Katharine Blodgett in the laboratory, 1938. Unidentified photographer. Source: Wikimedia.

Katharine Burr Blodgett (1898–1979) was an American physicist and chemist, born in Schenectady, New York. Her patent attorney father was murdered by a burglar shortly before she was born, and her mother moved Katharine and her brother to Europe, where they stayed mostly in France. Eventually they returned to the United States, and when she was 15, Katharine enrolled at Bryn Mawr College. She was 19 years old when a former colleague of her father, the future Nobel laureate Irving Langmuir (1881–1957), showed her the research laboratories of General Electric. He offered her a research position there upon the completion of her higher education.

In 1918, she received her master's degree from the University of Chicago and started to work with Langmuir, becoming the first woman scientist at General Electric. After six years of collaboration, Langmuir arranged for her to be accepted as Ernest Rutherford's doctoral student at the Cavendish Laboratory of Cambridge University. In 1926, she received the first PhD degree ever granted to a woman by Cambridge.

She returned to General Electric as a research scientist, working with Langmuir. They produced single-molecular layer-thin films on the surfaces of water, glass, and metal. These coatings were oily and extremely thin. The apparatus they used was called the Langmuir-Blodgett trough. Blodgett developed many practical uses for Langmuir's films. One of them was the nonreflecting coating that is called the Langmuir-Blodgett film, especially helpful in cinematography. The first major success of its application was for the film classic *Gone with the Wind* in 1939. She continued improving the technology. She dipped a metal plate into water that was covered by a layer of oil, then she dipped the plate in the water many times, adding oil layers onto the plate each time. The technology improved by controlling the thickness of the film, until she got an "invisible" glass.

During World War II, Blodgett and Langmuir worked on the development of protective smoke screens, submarine periscopes, and spy cameras. She invented a technique for de-icing airplane wings, allowing pilots to brave the weather under conditions that before had been too dangerous to fly. During her career, she was awarded eight U.S. patents and published 30 technical papers. Although her ideas and results were very important, she always stayed in the background. In his Nobel lecture for the 1932 Prize, however, Langmuir mentioned his joint work with Blodgett. Some of their joint work figures in the very short List of References in the printed version of that presentation.

Pioneers in Aviation and Space Travel

Women participated in the dawn of aviation, and two of the best known pilots are introduced here. I also present the first women space travelers.

Amelia M. Earhart. Source: Wikimedia.

Artist's rendering of Amelia Earhart in pilot's gear. It is part of an ornate gate by Brinsley Tyrell at Public School 291, 2195 Andrews Avenue North, University Heights in the Bronx, New York. Photograph by M. Hargittai.

Amelia M. Earhart (1897–1937) was an American aviation pioneer. She was born in Atchison, Kansas, and saw an airplane for the first time when she was 10 years old; by 23, she had piloted her first flight. She was the 16th woman in the United States to acquire a pilot's license. Soon she started establishing flight records and was the first woman to cross the Atlantic Ocean solo. In 1932, she took off from Newfoundland and landed in Northern Ireland. She earned a large number of decorations not only in America but also internationally. In 1937, she decided to circumnavigate the globe, again solo. For the last leg of the flight, she left from Papua New Guinea and disappeared somewhere over the central Pacific Ocean, near Howland Island. Her plane and body were never recovered. She has a memorial on the east coast of Papua New Guinea.

Amy Johnson in pilot's gear. Source: Wikimedia.

Amy Johnson's memorial plaque, Vernon Court, Hendon Way, London NW2.
Photograph by and courtesy of Steve Roffey.

Amy Johnson (1903–1941), a British aviator, studied at Sheffield University. In 1927, she moved to London, acquired a pilot's license, and, right away started establishing world records. In 1930, she became the first woman to fly solo from England to Australia. She served as a pilot in the Air Transportation Auxiliary during World War II. In January 1941, she was to deliver a Royal Air Force plane from Blackpool Airport to Kidlington Airbase in Oxfordshire, but she ditched off the Kent coast over the Thames in poor weather. Her body was never recovered.

A bust of Valentina V. Tereshkova in Ostankino Park, Moscow. Photograph by M. Hargittai.

Portrait of Valentina V. Tereshkova, 1963. Source: Wikipedia.

Valentina V. Tereshkova (b. 1937) was the first woman in a space flight, paving the way for an increasing number of female participants in the Soviet space program. The Russian word "cosmonaut" and the American word "astronaut" have the same meaning. Thus, Tereshkova was a cosmonaut. She orbited the earth solo in a Vostok 6 spacecraft in 1963. This followed soon after the first ever space flight by Yury Gagarin in 1961. Tereshkova graduated as an engineer, eventually reached a high rank in the Soviet Air Force, and became a high-ranking state official and member of the Duma, the Russian legislature.

Svetlana E. Savitskaya (b. 1948) was the second female space traveler. She participated in a space flight in 1982, and in 1984 she was the first woman to walk in space. She was a test pilot in her career and retired as an aviator.

In 1983 **Sally K. Ride** (1951–2012) became the first American woman in space as a member of the crew of the Space Shuttle *Challenger*. She was a physics PhD and a professor of physics at the University of California at San Diego. Besides her scholarly activities, she wrote science books for children. She was

A bust of Svetlana E. Savitskaya in Ostankino Park, Moscow. Photograph by M. Hargittai.

61 when she died of pancreatic cancer. A tree was planted in her memory at NASA's Johnson Space Center Memorial Tree Grove and a modest memorial tablet was placed next to this tree.

Judith A. Resnik (1949–1986) was the second American woman in space. She was an electrical engineer, software engineer, biomedical engineer, and a pilot. Her life was cut short in 1986 when the Space Shuttle *Challenger* exploded shortly after launch from the Kennedy Space Center.

A Soviet postage stamp showing Svetlana E. Savitskaya in cosmonaut's gear with two male cosmonauts.

Sally K. Ride in a NASA photograph.

An image of Sally K. Ride as part of an ornate gate by Brinsley Tyrell, at Public School 291, 2195 Andrews Avenue North in the Bronx, New York. Photograph by M. Hargittai.

NASA photograph of Judith A. Resnik.

NASA photograph of Kathryn D. Sullivan.

Kathryn D. Sullivan (b. 1951) is a geologist who became the first American woman to walk in space, in 1984, second after Savitskaya, also in 1984. Sullivan was a high government official in the Obama administration. When she retired, she was appointed a chair at the Smithsonian Institution.

Frances H. Arnold

Frances H. Arnold (b. 1956) traveled a long way before finding her calling. Her curiosity and dedication brought her to a unique research area and achievements honored by a Nobel Prize. She has successfully combined engineering and biochemistry and mimicked the process of evolution by artificial mutation of proteins, among them enzymes, that serve useful purposes.

She grew up near Pittsburgh, Pennsylvania. From early childhood she had broad interests and was a voracious reader who chose Christiaan Barnard, famous for performing the first heart transplant, for a role model. She was highly independent from an early age. When she wanted something she did not wait for others to provide it; she went after it herself, and usually got it. She attended local high schools wherever her family relocated. She graduated in 1974, already driven by social and political issues, attending protest rallies

Frances Arnold at Caltech, 2021. Photograph by Christopher Michel. Source: Creative Commons, https://commons.wikimedia.org/wiki/File:Frances_Arnold_in_2021_at_Caltech_01_(cropped).jpg.

against the Vietnam War. To earn money, she took diverse jobs and collected more experiences than some don't have over an entire lifetime.

She was determined to enroll at Princeton University, where her father received his PhD in experimental physics. Choosing her major was of secondary importance to simply getting in, which she did, graduating with a bachelor's degree in mechanical and aerospace engineering in 1979. It was a lonely place for a female student. Princeton did not have a long tradition of welcoming female students, especially not in engineering. She attended lectures in languages and economics to augment her education and felt ready to take on the world in whatever career she might land. There was one area that had special appeal to her, the environment and energy, specifically environmentally sustainable energy production. Her mentors at Princeton inculcated in her the "passion for connecting science and benefits to society." She channeled this passion into her "lifelong interest in alternative energy."[2]

A few detours followed in which she was piling up experiences before she went to graduate school at the University of California, Berkeley in 1981. Berkeley is unique in higher education for more than one reason; among them is that it has a separate School of Chemistry with a Department of Chemistry and an entirely separate Department of Chemical Engineering. Arnold chose the latter, with an emphasis on biochemistry. She missed chemistry in her previous studies, so she had a great deal of catching up to do. A host of new experiences enriched her existing ones: besides combining engineering and biochemistry, what it means to be a university professor and to establish spinoff companies. She graduated with a PhD in 1985. It was around that time that she met a Caltech professor of biochemical engineering, Jay Bailey (1944–2001). They married in 1987. Bailey's specialty was metabolic engineering. The cell produces certain substances, and Bailey's goal was to enhance such production by genetic and regulatory means, within the cell. Arnold's later research directions were influenced by Bailey's pioneering discoveries. He was at Caltech from 1980 until 1992, when he became professor of biotechnology at the Swiss Federal Institute of Technology Zurich, and worked there until a devastating cancer cut short his life. Arnold and Bailey had a son but separated around the time Bailey moved to Switzerland.

Around that time, Arnold met a cosmologist, Andrew Lang (1957–2010). They became partners and had two sons, but difficult days were ahead. Tragically, one of their sons died in 2016 in an accident. Lang was a successful scientist but prone to depression and eventually committed suicide. In 2004, Arnold was diagnosed with breast cancer, which had already spread to her lymph nodes. Surgeries, chemotherapy, and radiation treatment followed, and she persevered.

Arnold's career at Caltech began with a visiting position, which changed to a tenure-track assistant professorship. She was already a full professor in 1996 and had a named professorship in 2000. In 2017, she was appointed to the Linus Pauling Chair of Chemical Engineering, Bioengineering and Biochemistry. This professorship is a special honor; it was established in 1990, and Ahmed Zewail was its first holder, then 44 years old and a future Nobel laureate. Arnold is now the director of the Donna and Benjamin M. Rosen

[2] Frances H. Arnold, "Biographical," Nobel Prize, April 6, 2021, https://www.nobelprize.org/prizes/chemistry/2018/arnold/biographical/.

Bioengineering Center of Caltech. In January 2021, she was named co-chair of the President's Council of Advisors on Science and Technology. She has also initiated biotech companies and has been named to directorships of companies, among them Alphabet, Inc., the parent company of Google.

Suffice it to say that Arnold's science took off spectacularly at Caltech. The 1990s and the past decade (2011-) were outstanding periods in her research, with a growing number of young, talented, and dedicated disciples. She was determined to combine engineering and protein biochemistry, and the result was what is now well known as "directed evolution." Coining such phrases—and this one is excellent—is important in science because the name ties the discovery to a person more strongly than anything else.

First Arnold had to learn how to determine the sequence of amino acids in proteins. Then she had to find the technology to alter that sequence, which is a mutant sequence. Mastering that was not merely an academic achievement—and she started styling herself a protein engineer. The goal was to produce proteins of desired useful function. This is closely related to combinatorial chemistry, but directed evolution is aiming at producing proteins selectively rather than a complete library of proteins. They can be enzymes that are crucial components of the life cycle as catalysts, and as catalysts they could be employed in producing needed substances even at an industrial scale. Arnold and her laboratory make new enzymes to initiate new chemistries and to catalyze reactions that heretofore have not been known in the biological world.

The 2018 Nobel Prize in Chemistry amplified her tremendous success. She received half of the prize "for the directed evolution of enzymes."[3] Of the substances produced with her new enzymes, the Nobel Committee singled out pharmaceuticals and renewable fuels and the fact that their manufacturing is accomplished in an environmentally friendly manner.

Arnold is a rare, perhaps so far a unique female recipient of distinctions that are usually accorded to male engineers. I mention two: the Charles Draper Prize (2011), the highest honor for engineers in the United States, and the U.S. National Medal of Technology and Innovation, presented by President Barack Obama (2013). Arnold traveled a long way from getting into her father's alma mater to becoming a world-renowned protein engineer.

[3] The other half was shared by George P. Smith and Sir Gregory P. Winter "for the phage display of peptides and antibodies."

9

Ecologists

Ecology is a broad-based domain of science. It includes the study, protection, and preservation of our environment. It extends to biology, chemistry, and everything else that influences the relationship between humans and their surroundings, including when the products of nature are utilized for curing diseases. Female scholars and thinkers, as well as activists, have been at the forefront of ecology for decades.

Rachel Carson

Rachel Carson (1907–1964) started her career as a science writer and combined her literary acumen with scholarship. She observed the damage on the environment of the overuse of DDT and other pesticides and published *Silent Spring*, which has become a touchstone for the modern environmental movement.

Rachel Carson on a U.S. postage stamp.

Rachel Carson's pavement marker on the Extra Mile in Washington, D.C. Photograph by M. Hargittai.

She was born near Springdale, Pennsylvania, on the family farm, which provided a closeness with nature. She attended good schools and was invariably at the top of her class, whether in high school or when she sat for the civil service exam. She attended Johns Hopkins University, majored in zoology and genetics, but, at the time of the Great Depression, could not go for her doctorate because of family obligations. She found employment with the U.S. Bureau of Fisheries, where she worked in a friendly environment. She wrote for educational broadcasts and was promoted to the position of aquatic biologist. She published a trilogy on sea life, the volumes coming out in 1941, 1951, and 1955, each with a different, though excellent publisher.

Following World War II and the rise of science in warfare and national defense during the Cold War, there was an overwhelming tendency to solve the most diverse problems with technology. Such was the case with the use of pesticides, DDT most prominent among them. Carson was uniquely qualified as a scholar to evaluate the utility and potential harm of the uncontrolled use of pesticides and as a writer to present her position in a convincing way. Her book *Silent Spring* has become a classic. The title refers to the deafening silence she observed during spring time because bird life had suffered from the indiscriminate use of pesticides. Loss of birds was one of myriad potential dangers, alongside human cancer. Originally, "Silent Spring" was to be the title of one of the chapters of the book, the one presenting the impacts of

DDT and other chemicals on bird life. It was so succinct and poetic, though, that it became the title of the book.

Of course, she was not the only one to notice the adverse effects of the overuse of pesticides on the environment, but her book was to become the most efficient tool in the hands of the rapidly growing environmental movement. She anticipated that there would be much resistance to the advocacy represented by her book and was in no hurry to finish it. She wanted it to be well founded in its arguments and as complete as possible in its presentation, which considered not only short-term harms but also long-range hazards. One of them was the development of pesticide resistance by the targeted pests. This would weaken the ecosystem, and invasive species could then spread uncontrolled in the environment. Such resistance could be developed among mosquitoes, which could have tragic consequences for the fight against malaria.

As anticipated, there were strong forces condemning *Silent Spring* and Carson's activities. The loudest and most powerful among them were the chemical industries. They had a vested interest in ramping up pesticide uses. But Carson was precise in her reasoning; she did not want to eliminate the use of pesticides outright, only to stop their overuse and indiscriminate application. She stood out as a physically fragile but strong-willed individual who single-handedly fought against powerful adversaries. Her activities stopped the tide of indiscriminate use of pesticides and served as a pivotal marker for the environmental movement. She was a heroine who could readily be compared to another heroine, Frances Kelsey of the Food and Drug Administration (chapter 6), whose near-hopeless struggle prevented the introduction of a most harmful drug, thalidomide, for treating the morning sickness of pregnant women.

The heroic nature of Carson's fight for the environment can be appreciated even more if we consider that before, during, and after the publication of *Silent Spring*, she was fighting her own battle against breast cancer. She finally succumbed to this devastating illness, but in her larger goal, protecting the environment, she triumphed. The benefits of her resounding victory reverberate to this day.

Miriam Rothschild

When Miriam Louisa Rothschild (1908–2005) passed away, all major British newspapers carried her obituary. Her family was famous, but she was an

Miriam Rothschild in her home, 2002. Photograph by M. Hargittai.

A bust of Miriam Rothschild at the headquarters of the Royal Society (London). Photograph by M. Hargittai.

exceptional person in her own right. She was called a "high-spirited naturalist," "Entomologist Extraordinaire," "Queen Bee," "Lady Flea," and a lover of and fighter for all living things. Although she did not have much formal education or scientific training, she received honorary doctorates from eight universities, among them Oxford in 1968 and Cambridge in 1999. She was elected Fellow of the Royal Society in 1985 and was knighted for her achievements in 1999.

In 2002, my husband and I visited her on the Rothschild family estate, where she was born and lived most of her life. She was 93 years old, still full of energy and plans for the future. It was an exceptional experience to talk with her; she was a wonderful host, pleasant, interested, and she enjoyed talking about her life as if reliving it all over. She was speedily crisscrossing around in her wheelchair, so we hardly noticed her incapacitation.

She was born in Ashton-Wold, near Peterborough, in Northamptonshire, into the English branch of the Rothschilds as the first of four children. Among her earliest memories were the field trips when they were busy catching butterflies and ladybugs. These experiences triggered her interest in entomology, which shaped her life. Her father and uncle were amateur zoologists and entomologists. Another uncle, Lord Lionel Walter Rothschild, was the head of the N. M. Rothschild & Sons Bank, but his passion was zoology, which he readily shared with Miriam.

She enrolled at the University of London to study zoology and English literature. Already before the war she had tried to make arrangements with the British government to allow German Jews to come to England. Eventually, she brought out 49 Jewish children from Germany, between the ages of 9 and 14, and gave them a home at Ashton-Wold. During the first two years of the war, she worked at Bletchley Park with other biologists, mathematicians, and philosophers under Alan Turing on the famous Enigma project decoding German communications.

During this period, she met her future husband, George Lane, born György Lányi in Hungary. He became an officer in the British Army and was engaged in cross-Channel intelligence-gathering operations. They married in 1943, had four children and adopted two more. In 1957, they divorced but remained friends.

In 1947, Collins Publishers asked her to write a popular book on parasites. She was expecting her second child, and writing seemed an ideal occupation. However, she was doubtful such an unappealing topic could find much of an audience. Would people find it pleasing to read about the life cycle of a worm, *Halipegus,* that gets from the water of a pond into the liver of a snail, then into the cavity in a shrimp's body, into the gut of a dragonfly larva, and finally under the tongue of a frog? Would anyone find appealing the story of a worm that lives under the eyelid of a hippopotamus and feeds on its tears? To learn that fleas have the most complicated penis on earth? But she herself was enthusiastic, and the publisher was persuasive. *Fleas, Flukes and Cuckoos: A Study of Bird Parasites* appeared in 1952 and was a success.

She and George Hopkins cataloged the fleas that her father collected; this became a monumental five-volume treatise and took three decades to complete, all while she was raising six children. She usually worked on the book during the night—luckily for her writing, she had insomnia. The book series catalogs 30,000 specimens and made her the world authority on fleas. She studied their behavior and made several discoveries. Perhaps the most remarkable of these concerns the jump of the flea. She and her colleagues studied the jumping mechanism of the rabbit flea with the help of high-speed photography. The height of the jump of a flea compared to its body size is proportionally equivalent to a human jumping as high as the top of the Empire State Building. And the acceleration is enormous—20 times the acceleration of a moon rocket reentering the earth's atmosphere. Some fleas can jump as many as 30,000 times without stopping.

During the war, she was asked to look into the problem of tuberculosis among cattle in England. She investigated hundreds of tissue sections by microscope and determined that the transmitters of the disease were the wood pigeons with darkened plumage that carried tuberculosis in their adrenal glands. Another of her amazing findings was that the reproductive cycle of the rabbit flea is adjusted to that of its host. The rabbit flea can time its fertility so precisely that the baby fleas can drop right onto newborn rabbits. The general observation followed: the reproductive cycle of an insect parasite depends on its host. In the 1950s, there was an outbreak of the disease myxomatosis in Britain. She and her colleagues determined that the rabbit flea was its carrier and not the mosquitoes, as was previously supposed. Several of her papers discussed the defense mechanism of insects. Her account, coauthored with the Nobel laureate Tadeus Reichstein (1897–1996), became one of her most cited publications.

When we visited her, she was still active in research. She had a well-equipped laboratory at Ashton-Wold, and scientists from all over the world came there to collaborate with her. She participated in countless social and civic activities. She started a movement to save English wildflowers and cultivated them herself. Many people, among them Prince Charles, applied her ideas in their gardens. She created a Schizophrenia Research Fund to promote treatment. As early as the 1950s, she stood up for the rights of homosexuals, claiming that a woman bringing up six children in a happy family is in an ideally objective position to speak on this topic. She fought for animal rights, for better conditions in slaughterhouses, and against the misuse of animals.

Although she published over 350 scientific papers she did not consider herself a scientist. Rather, she regarded herself as the last of the old naturalists, a leftover species from the 19th century. Miriam Rothschild would have been famous just because of her family, but that kind of fame meant nothing to her. She was an idealist bursting with practicality. She wanted to make a difference, whether it was for sentimental or practical reasons. She fought against all kinds of injustice and wrongdoing, whether against humans or against the earth's fauna and flora. "I was always a tilter at windmills," she characterized herself.

Ayhan Ulubelen

Ayhan Ulubelen (1931–2020) was a pioneer in scientific research in Turkey and an internationally renowned scholar of natural products chemistry. She worked most of her life under much poorer conditions than most of her colleagues in the West.

She was born in Istanbul and originally thought of becoming a journalist, but during her high school years she saw a movie about Madame Curie and decided to become a chemist instead. She studied at Istanbul University and received her PhD from the Department of Pharmacy. For postdoctoral training she went to the University of Minnesota. She returned to Turkey and started her professorial career at Istanbul University, where she remained until her retirement in 1998, apart from a few years spent in the United States, Germany, and Japan as a visiting scientist. After retirement, she continued her research.

For centuries, there has been interest in the plants used in folk medicine for treating various ailments. There are special stores in Turkey, called *aktar*, that sell plants and plant extracts used as traditional medicine. Ulubelen and her colleagues identified and isolated their active ingredients and determined their structures by a variety of physical and chemical techniques. Some of these ingredients have been used against cancer, others against AIDS, yet others in cases of cardiovascular diseases, diabetes, and so on. The

Ayhan Ulubelen on the cover of *The Chemical Intelligencer* (April 1996). The portrait and the image of Istanbul University are by Istvan Hargittai.

investigation of plants that are potential remedies against cancer has long been a top priority. Ulubelen's group joined the plant-screening program launched in the early 1960s by the U.S. NIH to find cures for cancer. In this program, about 100 Turkish plants have been tested. *Ruta chalepensis* had for centuries been used by pregnant women to cause spontaneous abortion. Ulubelen's team checked its roots and aerial parts, isolated different compounds, observed their effects on mice, and identified several compounds of abortive activity. However, the follow-up studies showed that some of the mice developed cysts in their ovaries and had other problems as well. Therefore, they recommended against the use of this hazardous plant as an abortive agent.

Ayhan Ulubelen published over 300 research papers and received many awards and distinctions. She was a member of the NATO Scientific Committee for four years and a member of the Turkish Academy of Sciences. On November 2, 2011, 74 of the Academy's 137 members resigned, Ulubelen among them, in protest against a government decision to appoint two-thirds of the academy membership. In a few weeks' time, 17 former members of the Turkish Academy of Sciences—including Ulubelen—founded an independent, self-governing, civil-society organization to promote scientific merit, freedom, and integrity. Her social responsibility, in addition to her discoveries in science, is her legacy.

Chulabhorn Mahidol

The King and I and, later, *Anna and the King* are among my favorite movies. I could not help feeling that parts of these movies came alive when, in 1999, we visited the Princess of Thailand in Bangkok. Her full name was Professor Dr. Her Royal Highness Princess Chulabhorn Mahidol. She was the youngest daughter of Their Majesties King Bhumibol Adulyadej and Queen Sirikit of Thailand. Now she is the younger sister of the sovereign, King Vajiralongkorn. You might wonder how a royal princess gets into a book on women in science. The answer is simple: she *is* a woman of science, a professor of chemistry at Mahidol University and the founding president of the Chulabhorn Research Institute.

She may be the only princess chemist. Originally she wanted to become a concert pianist, but the king insisted that all his children learn a profession useful for the future of a developing country. First, she studied at Kasetsart University, and then went on to earn her PhD at Mahidol University, both in Bangkok. She was a postdoc in genetic engineering at the University of Ulm and at Tokyo University Medical School.

After returning home, she founded the Chulabhorn Research Institute, whose stated goal is to "improve the quality of life." Her interest and the main research line of the Institute is natural products chemistry. This is a fitting topic for Thailand, where there is a large arsenal of plants that have been used

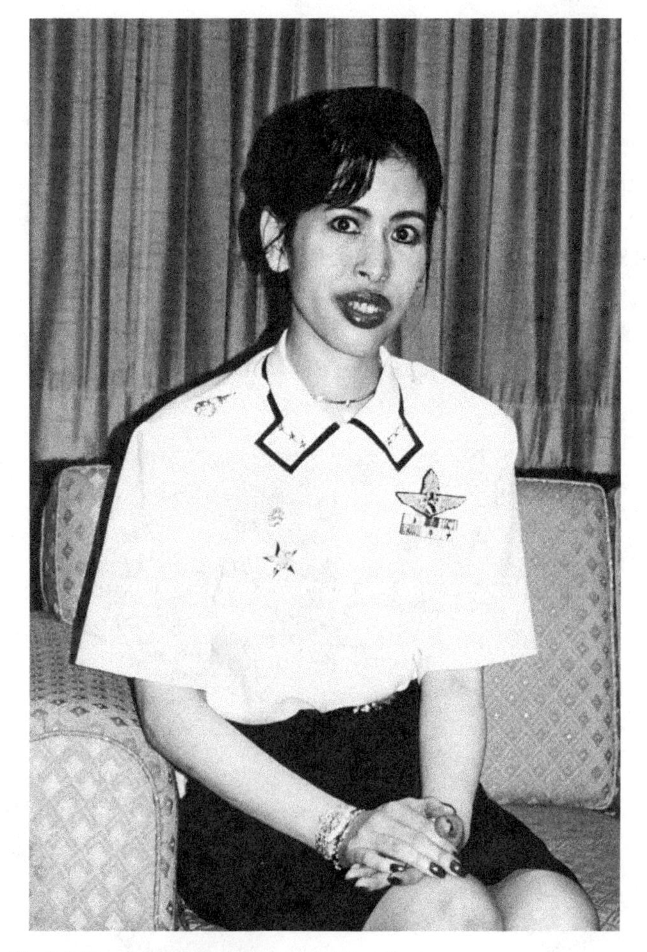

Princess Chulabhorn. Photograph by M. Hargittai.

for centuries to heal people. The princess was always fascinated when old people told her about the healing effects of various plants. There are thousands and thousands of different species of plants in Thailand. Her institute is well equipped with modern instruments, mostly gifts from Germany and Japan.

Besides teaching at the university, Princess Chulabhorn is an admiral, a general, and an air chief marshall. She teaches chemical warfare to personnel of the Thai Air Force. She arrived at our meeting from just such a lecture wearing her Air Force uniform. She stressed that she wanted her students to learn about the relevant chemicals, the biological agents, and about how to protect their lives, but not about how to kill.

Because of her many duties in research and education, she was excused from attending most ceremonial functions. Her role as a scientist is more important than her role as a princess. She is divorced and has two daughters (born in 1982 and 1984). For her active involvement in promoting scientific collaboration in Asia and the Pacific, she was awarded the Einstein Medal of UNESCO. She is the first person from Asia to become an Honorary Fellow of the Royal Society of Chemistry in England.

Jane Morris Goodall and Other Primatologists

Jane Morris Goodall. Source: Wikimedia, https://commons.wikimedia.org/wiki/File:Jane_Goodall_at_TEDGlobal_2007-cropped.jpg.

Jane Morris Goodall (b. 1934), born in London, is a primatologist and anthropologist, best known as an expert on chimpanzees, as the founder of the Jane Goodall Institute and the Roots & Shoots Program, and for her work on conservation and animal welfare. She is famous for her unconventional methods; for example, she "humanized" her animals, distinguished them by names and referred to them using terms that are applied to humans. In 1993 she wrote, "When, in the early 1960s, I brazenly used such words as 'childhood,' 'adolescence,' 'motivation,' 'excitement,' and 'mood,' I was much criticized. Even worse was my crime of suggesting that chimpanzees had 'personalities.' I was ascribing human characteristics to nonhuman animals and was thus guilty of that worst of ethological sins, anthropomorphism, instead of the then accepted method of numbering them."[1] She traces her interest and love of chimpanzees to having received a stuffed chimpanzee as a child, which she named Jubilee. It became her favorite toy, just as a teddy bear was for other children. She loved animals and the outdoors and was interested in Africa. During her first visit there in 1957, the Kenyan paleontologist Louis Leakey (1903–1972) sparked her interest in chimpanzees, and she realized that further studies would be needed if she was to investigate them seriously. She returned to England and studied primate behavior with Osman Hill and primate anatomy with John Napier, both leading authorities in these fields. In 1960, she moved to the Gombe Stream National Park in Tanzania. Her mother went with her, as was required by those running the Park. Indeed, it was rather unusual for a young lady to embark on the kind of project that Goodall had in mind. This was a welcome assignment for her mother, who very much encouraged her daughter's aspirations. Goodall became a trailblazer in a heavily male-dominated field. Today men and women work in this area in about equal numbers.

She has done a great deal of research and made important discoveries over the decades. It used to be an accepted view that only humans were capable of making tools, but she found that chimpanzees are also toolmakers, however rudimentary their tools may be. She also found that chimpanzees were not vegetarians, as was believed; they hunted and ate smaller primates. She invigorated the entire field of primate studies, established the Jane Goodall Institute, and has successfully campaigned against using animals in medical experimentation, zoos, farming, and sports. She has been active in

[1] Jane Goodall, quoted in Paola Cavalieri, ed., *The Great Ape Project: Equality beyond Humanity* (London: Fourth Estate, 1993), 10.

supporting "green" politics and politicians and has published and lectured extensively, for which she has been internationally decorated. She has also been involved in various controversies—unsurprisingly, considering the breadth of her activities. Perhaps her greatest achievement has been raising the level of awareness concerning habitat preservation.

I mention two other women primatologists who received much inspiration from Louis Leakey and, subsequently, from Jane Goodall. The main interest of **Dian Fossey** (1932–1985), the former California occupational therapist, was in mountain gorillas. The Lithuanian Canadian scientist **Birutė Galdikas** (b. 1946) has worked with orangutans. Goodall, Fossey, and Galdikas have been called "The Trimates," and sometimes referred to as "Leakey's Angels."

Youyou Tu

Youyou Tu, 2015. Photograph by Bengt Nyman. Source: Creative Commons.

Youyou Tu (b. 1930) was born in Ningbo, Zhejiang, China. She had four brothers, and her parents placed great emphasis on the education of all of their children, not only the boys. Tu attended good schools, but a tuberculosis infection interrupted her high school studies for two years. This defining experience moved her to become a medical doctor. Between 1951 and 1955 she attended Peking University Medical School/Beijing Medical College and became closely associated with its Department of Pharmacy. The instructions were broad-based, and a number of professors were so-called returnees from the West. She majored in pharmacognosy—the science of plants as natural sources of drugs—learning about the medicinal plants, their origin, classification, and identification, about their botanical description, and about phytochemistry. Students learned how to extract the active ingredients from the plants and how to determine their chemical structures. They also learned traditional Chinese medicine.

In the 1950s, government policy advocated a combined approach of Western and traditional Chinese medicines. This was in part because of the shortage of trained medical personnel and in part out of reverence for tradition. Even the personnel versed in Western medicine had to become acquainted with traditional Chinese medicine. Tu worked on researching Chinese plants for possible applications and participated in several projects from the start of her research career. Sometime in the second half of the 1960s, she was appointed to head a task force aimed specifically at finding a natural remedy for malaria.

Malaria has been a life-threatening epidemic disease, which for long decades was kept under control by chloroquine and quinolones. Then, in the late 1960s, drug-resistant versions of malaria appeared and were seemingly unstoppable. Once again, malaria became a global issue. In Vietnam, the U.S. military found it a great challenge. In China, it was also the military that showed a particularly keen interest in finding a cure. They started the search in 1964, initially in secret. In 1967, they set up National Office 523 to coordinate their actions. They screened thousands of substances without success. In 1969, the project involved the Academy of Traditional Chinese Medicine and the Institute of Chinese Materia Medica in the quest for novel remedies against malaria. It was at that point that Tu received her assignment and started building up what the government called Project 523. She was 39 years old, and thus considered very young for such a leadership role in China. Her family circumstances were difficult. Her husband was away at a "training camp"—it was the time of the so-called Cultural Revolution, a dark period in Chinese history. Scientists were distrusted, and Tu's new position brought all kinds of dangers with it. But it appeared to be in the interest of national security to solve this problem, so likely for this reason she was spared. In any case, Tu had to leave behind her two daughters; fortunately, she could rely on her parents to take them. Project 523 occupied her for the next three years. When she finally reemerged for the family, her younger daughter did not recognize her.

In the framework of Project 523, Tu, along with a team of able associates and assistants, investigated the history of traditional Chinese medicine. She interviewed all available medical practitioners and collected 2,000 prescriptions of herbal, animal, and mineral origin. In three months from the initiation of the project, she summarized 640 prescriptions in a booklet which contained all possible and hopeful candidates as remedies against malaria.

The promising candidate substances were tested on rodents infected with malaria. The team went back and forth between failing experiments and the literature. Finally, one compound showed some promise for intermittent fever, which is a hallmark of malaria. The herb was Qinghao, its Chinese name, in the *Artemisia* family of plants. This observation was strengthened by its mentions in the literature as a remedy for malaria symptoms. Still, many experiments were needed before the most efficient approach was found. For example, it turned out that the ingredients had to be handled at room temperature rather than boiling them out. Also, they found that the leaves of the plant contained the most efficient remedy in the highest concentration.

By March 1972 everything was ready for increased production of the substance labeled "191." Production capacities were very limited, as the Cultural Revolution had led to the shutdown of most pharmaceutical workshops. Tu and her coworkers had to work with rudimentary equipment under unsafe conditions, such as handling large amounts of organic solvents without proper ventilation. Because of some conflicting information from animal toxicological studies, volunteers were needed to test the various preparations on humans. Tu and some of her closest associates volunteered. The clinical trials commenced in 1972 and continued in 1973. The best candidate, the most effective against malaria, was named Artemisinin; its Chinese name was Qinghaosu. After its chemical structure was determined, Tu tried to see if its chemical derivatives might be as good or even more effective. Eventually she arrived at a high-efficacy derivative, Dihydroartemisinin. It proved 10 times more potent than Artemisinin itself. An enormous number of tests and clinical trials followed. Finally, in 1986, the Ministry of Health granted the necessary certificate for new drugs for Artemisinin and, in 1992, for Dihydroartemisinin. Both showed "high efficacy, rapid action and low toxicity."[2]

This success story demonstrates the value of both traditional folk medicine and the perseverance of individual researchers, even when it comes at severe personal sacrifice. It also shows that success can be achieved under unfavorable conditions, even in spite of those conditions. For Tu, once success arrived, appreciation and rewards followed. But this was neither straightforward nor fast. Tu published her research in 1977, but anonymously. Only in 1981 could she step out of anonymity and report her findings at a meeting of the World Health Organization. The highest recognition came in 2015 with the Nobel Prize in Physiology or Medicine "for her discoveries concerning a novel therapy against Malaria."[3] It was followed by the highest awards in China in 2016 and 2019.

[2] Tu Youyou, "Biographical," Nobel Prize, April 7, 2021, https://www.nobelprize.org/prizes/medicine/2015/tu/biographical/.

[3] This was half of the 2015 prize. The other half went to William C. Campbell and Satoshi Omura "for their discoveries concerning a novel therapy against infections caused by roundworm parasites."

Name Index

For the benefit of digital users, indexed terms that span two pages (e.g., 52–53) may, on occasion, appear on only one of those pages.
Figures are indicated by *f* following the page number